BASICS OF ELECTROORGANIC SYNTHESIS

BASICS OF ELECTROORGANIC SYNTHESIS

Demetrios K. Kyriacou

The Dow Chemical Company
Western Division
Research and Development Department
Pittsburg, California

A WILEY-INTERSCIENCE PUBLICATION

JOHN WILEY & SONS   New York · Chichester · Brisbane · Toronto

Library of Congress Cataloging in Publication Data:

Kyriacou, Demetrios K
  Basics of electroorganic synthesis.

  "A Wiley Interscience publication."
  Bibliography: p.
  Includes index.
  1. Electrochemistry. 2. Chemistry, Organic—
Synthesis. I. Title.
QD273.K97 1981   547'.2   80-25326
ISBN 0-471-07975-8

Printed in the United States of America

10 9 8 7 6 5 4 3 2 1

To the memory of my Father

PREFACE

It is almost a standard requirement that the author of a new textbook prepare an apologia for writing the book. The plethora of such books apparently demands this formality. There is, however, no compelling reason to write one for this book. No other field of chemistry is represented by fewer books than electroorganic synthesis. Only a few such books have recently made their appearances. One can easily discern, in the prefaces of these books, the authors' passionate hopes that this neglected field of organic synthesis can be popularized. It is absurd that even advanced organic chemistry books find it unnecessary to spare a paragraph's space for electroorganic synthesis.

There is, therefore, a need for textbooks about electroorganic chemistry at all levels, and for books written especially for organic practitioners in industrial laboratories.

This textbook is written for the industrial organic chemist who would like to use the electrolytic method of synthesis but who lacks the background knowledge to do so. No attempt is made to document the status of electroorganic synthesis by quoting exhaustively from research papers in the vast international literature. That is unnecessary and entirely beyond my ability and the scope of this book. Rather I portray the *idiosyncracy* of the electroorganic phenomenon and its practical synthetic value. Selected examples from the literature and from my own work are described at a *beginner's level*.

The book consists of four short chapters and an appendix. Chapter 1 describes the general electroorganic reaction and illustrates the general mode of *thinking* and *doing* in the area of electrosynthesis. Chapter 2 is a brief survey of electroorganic reactions. Chapter 3 is concerned with some special topics, and Chapter 4 with the praxis of electroorganic synthesis. The Appendix contains some basic electrochemical concepts and principles, as well as various items and comments useful to the neophyte in the field. I have also included, in the General Bibliography, references to other introductory and advanced books on electrochemistry and electroorganic synthesis.

I would like to thank my wife, Eléni, who patiently typed the manuscript.

DEMETRIOS K. KYRIACOU

Pittsburg, California
March 1981

CONTENTS

BASICS OF
ELECTROORGANIC
SYNTHESIS

INTRODUCTION

In their distant origins all chemical phenomena reveal themselves as fundamentally electrical. All chemical reactions, therefore, might be called electrochemical reactions. From a practical perspective, however, an electrochemical reaction is one that can be performed in an *electrolytic cell.* The method of carrying out organic syntheses in electrolytic cells is the *electroorganic* method of synthesis.

All cells must contain at least two electrodes in contact with an electrically conducting liquid medium in order for the *electrodic,* or electrochemical, reactions to be possible. The term electrodic is very descriptive in that it implies the necessary presence of electrodes. Electrodes are conductors of electricity. They constitute one phase of the electrolytic system, while the liquid medium, or the solution to be electrolyzed, constitutes the other phase. The presence of these two phases clearly defines all electrodic reactions as heterogeneous reactions. The heterogeneous nature of the reactions is in the elementary event, namely, the *electron exchange act* at the electrified interface between electrode and solution. This electron transfer phenomenon is, of course, omnipresent in nature. Electrochemistry, therefore, is an interdisciplinary science. As such it could be rigorous in its demands on its student and practitioner. However, the knowledge of electrochemistry needed for most practical organic syntheses in the laboratory is very modest, while the benefits can be disproportionately great.

Electroorganic synthesis is an old method. Despite that, it has been practiced by only a small number of industrial chemists since the time of Faraday. One of the main reasons, most observers agree, for the virtual absence of electroorganic research laboratories in industry and for the very slow progress of industrial electroorganic processes has been the difficulty of scaling up such processes. Another reason might be that only a few organic chemists in industrial laboratories have had the opportunity to become knowledgeable in this chemicophysical, as it were, method of synthesis. In addition, only in the last two decades has an understanding of the fundamental mechanisms of electroorganic reactions become possible. The current scarcity of raw materials and energy has generated a vigorous interest in electroorganic synthesis. New and improved cells and electrode materials, and a better understanding of the underlying principles are bound to mitigate many of the scale-up problems.

In certain specialty areas, such as the pharmaceutical and agrochemical areas, where chemical and stereochemical selectivities are needed, and where the electricity cost might be a relatively minor factor for large-scale production, the electrolytic method of synthesis could be especially desirable. Most importantly, electroorganic chemistry can be called upon to contribute in the efforts to abate pollution. The bench organic chemist, for whom this book is written, will find the technique of electrosynthesis to be a very convenient alternative, as well as one with very high intellectual appeal.

1
ELECTROORGANIC SYNTHESIS AND TECHNIQUE IN GENERAL

The underlying primary physical process for all chemical and electrochemical reactions must involve the *motion of electrons.* In purely chemical reactions this motion creates the kind of activated complex that leads to products. In electrochemical reactions the motion of electrons is different. This chapter is an exposition of the nature of the electroorganic reaction insofar as it pertains to organic synthesis.

1.1a The Generalized Electroorganic Reaction

All electrochemical reactions are referred to as either *anodic* or *cathodic* reactions. The former occur at the *anode,* that is, the positive electrode, while the latter take place at the *cathode,* the negative electrode. Such reactions are carried out in a suitable *electrolysis cell,* or *electrolyzer,* as depicted in Fig. 1.1.

Anodic reactions are fundamentally *deelectronations,* or oxidations in the usual sense, whereby the substance gives up electrons to the anode, as it would to an oxidizing agent. Cathodic reactions are *electronations,* or reductions, whereby the substance accepts electrons from the cathode, as it would from a reducing agent.

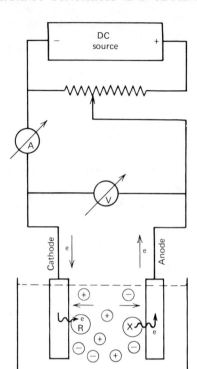

Figure 1.1   Schematic Representation of Electrolysis Cell.

As a result of these primary electron transfer reactions, chemical bonds break and new bonds are formed, as with all types of organic reactions. Almost all types of organic reactions are possible by the electrochemical method: additions, substitutions, cleavages, eliminations, couplings, cyclizations, oxidations, and reductions of functional groups. Our concepts of *polar* and *free-radical* mechanisms can be applied to electroorganic reactions in the same way as to conventional organic reactions.

The generalized electroorganic reaction is symbolized in Fig. 1.2. Note here that the electrode, the anode (Fig. 1.2a) is shown to be deficient in electrons near its surface. The organic species, R, *reduces* the electrode by donating electrons to it. These donated electrons are removed from the reduced surface via the external circuit, the wire, so that the reduction reaction may continue. Were it not for this electron removal, the electrodic reaction would stop as soon as electrostatic forces arose that opposed further *net* electron transfer at the interface between electrode and solution. For cathodic reactions (Fig. 1.2b) the events are similar but, in the physical sense, exactly oppo-

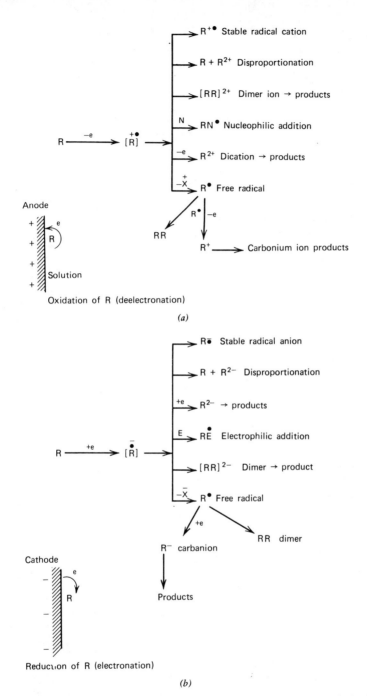

Figure 1.2 The generalized electroorganic reaction. (*a*) Anodic reaction. (*b*) Cathodic reaction.

site. (In the symbolism used a dot indicates an electron, while a dash stands for a pair of electrons and also a single negative charge. The terms electronation and deelectronation are adopted from Bockris and Reddy.[1]) The sequence of events in these schemes is designated ECE . . . , CEC . . . , and EEC . . . , where E and C denote an electron transfer step and a chemical step, respectively.

The fundamental event in electrodic reactions is the electron exchange at the electrode–solution interface. This electrified interface, or the *electrical double layer,* is a very narrow region under the influence of enormous electric fields (as high as 10^6 to 10^7 V/cm). It is the presence of such fields that differentiates electrochemical and conventional heterogeneous chemical or catalytic reactions. The organic chemist can readily discern both the intrinsic merits and demerits of the heterogeneous nature of the electrolytic method of synthesis. For instance, electroorganic reactions are expected to be generally slower than the homogeneous reactions, but they might be more selective chemically, and particularly stereochemically. In an electrochemical reaction the electric field and the *activity* of the electron as the primary reactant are capable of very easy experimental modification and control. Thus merely by manipulating a few knobs in a potentiostat or rectifier, it is possible to start a reaction, to control its rate, to change its course, and to stop and start it again at will. Such control of a reaction is rarely possible in conventional chemical (thermal) reactions or catalytic processes.

The practical essence of the electroorganic reaction consists of three basic steps:

1 Transfer of the organic species to the electrode surface.
2 Exchange of electrons between electrode and organic species. Adsorption may be involved here.
3 Removal of the primary products from the electrode surface (this is necessary for the reaction to continue). Desorption may be involved here.

The slowest of the three steps determines the overall reaction rate.

It may be worthwhile here, although it is somewhat precocious, to make a formal comparison of a purely chemical reaction with an electrochemical cell reaction. Consider the generalized chemical reaction:

$$A + B \rightleftharpoons [AB] \longrightarrow C + D$$

This reaction requires the close physical contact of A with B in order to take place. The rate of the reaction depends on the factor $e^{-E/RT}$. A change in the rate can be effected by a change in temperature. There is little that can be done with E, the activation energy for this reaction.

If the same reaction is to be accomplished electrochemically in an electrolytic cell, two *half-cell* reactions must take place:

$$A + e \rightleftharpoons [A \cdot e]^{-} \longrightarrow C \qquad \text{cathodic}$$
$$B - e \rightleftharpoons [B]^{+} \longrightarrow D \qquad \text{anodic}$$

The sum of these reactions is the overall cell reaction, which is identical with the chemical reaction:

$$A + B \longrightarrow C + D \qquad \text{overall cell reaction}$$

In the electrochemical cell reaction the reactants A and B do not have to meet as they must in the chemical reaction. The reactants in the cell exchange electrons *from a distance* by means of a metallic wire in the external circuit. The rate of the electrochemical reaction depends on the electrochemical activation energy E' in the factor $e^{-E'/RT}$. The chemical activation energy cannot be changed experimentally, whereas the electrochemical activation energy can be very conveniently changed by varying the voltage applied on the cell. The essence of both reactions is, of course, the same: *motion of electrons*. It is the physical and experimental mode of this motion of electrons that constitutes the difference between the chemical and electrochemical reactions.

1.1b   Reaction Variables in Electroorganic Synthesis

In general, electrochemical systems are richer in variables than the purely chemical systems. Because many of the variables are not well understood—and even if they were, they might not always be easy to apply and to control in plant-scale works—it is difficult for the electroorganic process to advance beyond the laboratory bench or the pilot-scale stage, in spite of the recent progress in understanding the fundamental mechanisms of electroorganic reactions.

Some reaction variables are common to both chemical and electrochemical reactions, although their effects in each type of reaction can be very different. Thus temperature, solvent (protic or aprotic), pH, concentrations of reactants, relative amounts, methods of mixing, and reaction time are common variables. In addition, electrochemical reactions involve electrical and other physical variables, the most important of which are:

1 The electrode potential, which can be controlled.
2 The electrical double layer. This variable is the least understood, and its effects are the most difficult to predict.
3 The electrode material and its physicochemical characteristics.
4 The concentration and kind of supporting electrolyte; this would affect the conductivity and the accessible potential range of the medium, the structure of the electrical double layer, and the nature of the final products.
5 The type of cell (divided or undivided) and the diaphragm, if divided.
6 The physical composition of the electrolysis medium (homogeneous or heterogeneous).
7 The degree and means of agitation.

It is correct to conclude that the greater the number of operating variables, the more difficult it is to control the course of a reaction. However, it might be argued that the more degrees of freedom, the greater, perhaps, the potential versatility of a system.

For every electroorganic synthesis in the laboratory, therefore, six principle questions must usually be considered:

1 The selection of the electrodes.
2 The selection of the reaction medium (solvents and supporting electrolyte).
3 The selection of the optimum (practical) electrode potential (and cell voltage).
4 The selection of the optimum temperature range and very often the pH of the medium.
5 The selection of the cell type (divided or undivided).
6 The recovery of the products.

Although the elementary step in all electrolytic reactions is the electron exchange at the electrified electrode–solution interface, the final products of the reaction in most cases will depend, to various degrees, on how well the questions above have been answered. Obviously, if the purpose of the electroorganic synthesis is to eventually scale up the method for commercial production, the thinking will have to be modified in accordance with the demands not only of *science* but also, and usually more, with those of technology and economics. This aspect needs no elaboration, since the organic chemist in industry is well aware of it.

1.2 Setting Up the Electrolysis Cell

1.2.1 The Basic Laboratory Apparatus

The basic components of an electrolysis apparatus are five: (1) the dc power supply; (2) the electrodes; (3) the vessel containing the solution; (4) the voltage measuring device; and (5) the current measuring device.

For aqueous or aqueous-organic media a power source of 10 to 20 V with current output capability up to 10 A would be sufficient for most laboratory-scale organic syntheses. For less conductive media of organic solvents a power source of up to 100 V would be needed. Generally, laboratory preparative electrolyses are carried out in cells of 50 to 500 ml capacity and with solution volumes of 10 to 300 ml.

1.2.2 Two-Electrode Cells

Many electrolytic preparations can be done in simple two-electrode cells. Such cells are represented by diagrams such as Fig. 1.3a. An ordinary glass beaker may serve as the electrolysis vessel. Two electrodes, one to function as anode or positive electrode and the other as cathode or negative electrode, and the solution to be electrolyzed are placed in the beaker (see Fig. 1.1 and cell designs in Section 1.3). The electrodes are connected, usually with copper wire connectors, to the dc power source. The present availability of potentiostats and rectifiers as dc sources has made the work of the electroorganic chemist pleasant and very easy. An ammeter and a voltmeter are part of the external circuit, as shown in Fig. 1.1. To minimize the resistance of the solution (*IR* drop in the solution and hence waste of electric energy as heat), the distance between the electrodes should be small, usually 0.5 to 5 cm, depending on the conductivity of the medium. The electrical connections of the wires should be firmly secured in order to avoid unnecessary voltage losses and the possibility of sparks (this is especially important when using flammable solvents).

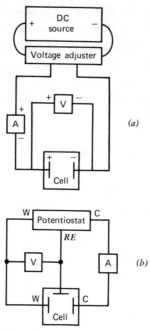

Figure 1.3 Block Diagrams of (*a*) two- and (*b*) three-electrode cells.

1.2.3 Three-Electrode Cells

It is generally desirable, and in a great number of cases necessary, to employ three-electrode cells. In such cells the third electrode is not intended to carry any current, that is, it does not take part in the cell reaction. It functions only as a *reference electrode* against which the potential of the *working electrode* is measured. The working electrode is the one at which the reaction of concern takes place. Three-electrode cells are depicted in Fig. 1.3b. The potential of the working electrode, or its reducing or oxidizing power, can be varied by varying the applied cell voltage, that is, the potential difference between anode and cathode. The more negative the electrode potential, the greater its reducing power, while the more positive the potential, the greater the oxidizing power. This, of course, is true with the two-electrode cell also, but in that case only the cell voltage is known. It is usually more important, however, to know the potential of the working electrode. It is the impression of a definite potential on the electrode that enables the electrons to cross the energy barrier between electrode and solution in sufficiently great numbers and thus brings about the desired electrochemical reactions.

In electrochemical reactions, the experimenter can raise or lower the electrode potential and fix it at the most favorable value, or range, for the desired synthesis. This can be done either manually or automatically. In the latter case *potentiostats* are now used. These instruments are very convenient and are commercially available at moderate cost. They operate electronically, converting ac to dc and maintaining the working electrode potential at a set value during electrolysis. They are amply described in the literature and in the manuals of the manufacturers.

For plant-scale operations rectifiers are employed. Even with rectifiers it is possible to maintain the potential of the working electrode within reasonable limits (ie, ± 0.2 V) if necessary, by means of the rectifier voltage settings. Usually, however, constant electrolysis conditions and cell geometry (continuous processes) can, in effect, maintain the working electrode potential at the desired value.

1.3   Laboratory Cell Designs

Laboratory cells for preparative electrolysis may have various designs. Some cells are very simple while others are complex depending on the type and purpose of the electrolysis. The experimenter is free to use all imaginative powers in designing laboratory cells. For illustrative purposes some very common and generally adequate cells are schematically shown in Fig. 1.4. The reader is referred to ref. 2 for a thorough and extensive description of all kinds of experimental cells (see also the General Bibliography).

The terms *static* and *flow* cells are used to differentiate the manner of operation. For static cells agitation is provided by either a mechanical or magnetic bar stirrer. Static cells are most often used in the laboratory. Flow cells require more equipment and are more suitable for pilot-scale or plant-scale works.

1.4   Selection of Electrode Material and Cell Geometry

1.4.1   The Electrode Material

The most important (and absolutely necessary) part of any electrolytic system is the electrode. In some cases, and especially in one-compartment cells, the nature of both anode and cathode may be of about equal importance. Usually, however, it is the working electrode that is of greatest concern. The most prudent way to select an electrode, that is, its material and its surface morphology, is, as a general rule, by direct experiment. Some help may be obtained from certain theoretical considerations and from a knowledge of electrodes

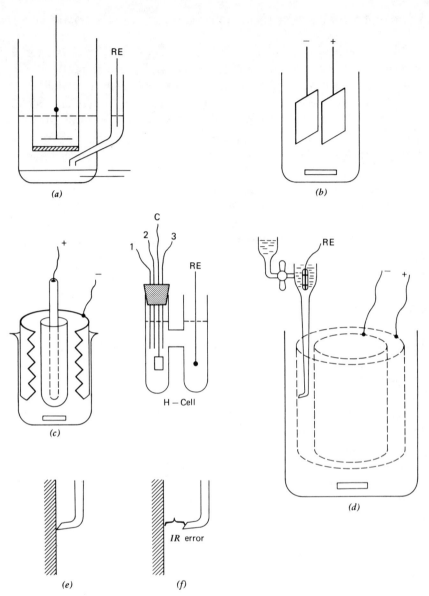

Figure 1.4 Common cells for bench-scale electrolysis. (*a*) Divided cell (fritted glass diaphragm, mercury pool cathode). (*b*) Undivided cell (parallel plate electrodes). (*c*) Divided cell (cylindrical symmetry electrodes, porous cup diaphragm). (*d*) Cylindrical screen electrodes, use of Luggin capillary. Anode is working electrode. (*e*) Proper positioning of Luggin capillary tip. (*f*) Error caused by *IR* drop. Cathodic potentials appear more cathodic and anodic potentials more anodic. H-Cell. Cell for rapid testing of electrodes (1, 2, 3) for a particular reaction. Currents of a few milliamperes are used with such a cell. RE: reference electrode.

that have already been tried. Different electrodes can best be evaluated by using the same electrolysis medium under similar conditions. An expedient and reliable preliminary evaluation of electrode performance can be made by polarographic and voltammetric methods. These methods furnish information about current-potential relationships in various electrolytic systems. Various electrodes are tried in the same solution. If an electrode, in a certain medium, shows no polarographic or voltammetric activity (see Section 1.7) it would most likely not perform under preparative-scale electrolysis.

Any electronic conductor may be considered as an electrode material. However, in practice, there are always limitations, especially for anode materials. Electrodes may be prepared from metals in the form of plates, rods, wires, gauzes, wool, sheets, mercury pools, and from other conductive materials such as graphite, carbon, carbon pastes, and various metallic oxides. Table 1.1 lists most of the electrode materials commonly used. Alloys of various metals may also be used, and thus the list might be much longer than shown here. The electrode material is of practical significance because it affects the electrochemical reaction, but so does the microscopic surface morphology of the electrode, more so in many cases. In fact it is the electrode's surface in contact with the solution that is the *actual* electrode. This is very

Table 1.1 Commonly Used Electrode Materials

Anodes	Cathodes
Platinum	Mercury
Graphite	Lead
Carbon	Tin
Lead dioxide	Zinc
Metals coated with	Cadmium
precious metals, eg,	Nickel
ruthenium, irridium,	Monel
rubidium	Chromium
Nickel oxide	Aluminum
Silver oxide	Copper
Manganese oxide	Iron
Iron oxide	Steel
Gold	Graphite
	Carbon
	Silver
	Platinum
	Gold

important to understand, because even traces of metallic impurities in a solvent may electroplate on the initial electrode surface and completely convert it, in effect, into a different electrode. Heterogeneous catalysis and *electrocatalysis* depend on the kind of atoms and their characteristic arrangement on the electrode surface.[1] It is of utmost importance, therefore, when an electrochemical reaction is possible only at a certain electrode, to ensure that the electrode surface remains the same during the initial electrolysis and in subsequent runs. Occasionally, electrodes behave well at the start of an electrolysis but lose their effectiveness in the course of the reaction or in subsequent runs. As an example, it was observed in the author's laboratory that spongy silver electrodes almost completely lost their activity when they were left to dry in air after an electrolysis run. They were reactivated by an anodization procedure. The phenomenon of an electrode's diminished performance or deactivation is known as *passivation*. A consistent procedure in the preparation and treatment of electrodes, before, during, and after use, may be necessary for reproducible results. Many electrosyntheses, however, are possible with electrodes that need no elaborate treatment or no treatment at all, other than the preservation of a conductive surface. For aqueous or aqueous-organic media it is possible, sometimes, to select electrodes on the basis of their *hydrogen overvoltage* characteristics (see the Appendix). For example, if the purpose of the electrolysis is to hydrogenate a substance with nascent hydrogen by the electrochemical method (electrohydrogenation), it is generally better to start experimenting with the *hard* metals, namely, platinum, nickel, iron, and copper, which have low hydrogen overvoltage, instead of with the *soft* metals, lead, cadmium, zinc, tin, and mercury, which are metals of high hydrogen overvoltage. On the other hand, if *direct* electrochemical reductions are to be effected, the *soft* metals would most likely be preferable. There is indeed the possibility of successfully combining some intuition, some art, and some science in this area of electrosynthesis. Very active electrodes may sometimes be made *in situ* by adding only a few tenths of a percent of a certain metal ion. Silver ion, for example, may be added to alkaline solutions. The fine silver oxide particles that form upon addition of a few milliliters of aqueous silver nitrate to the solution in the cell are rapidly reduced at the negative electrode to form a *spongy* coating on the electrode, which may be either silver or some base metal inactive by itself. Spongy, or high surface, electrodes to be used either as cathodes or as anodes can be made with various metals and in various media. Various plating and polarity reversal procedures can produce such active surfaces. The reader will find many specific examples in the literature on this subject.

When electrodes become passive to an undesirable degree, various chemical, mechanical, and electrical methods are used in restoring their activity. It should be understood that electrodes requiring very special care and treatments are impractical for industrial synthetic processes.

1.4.2 Cell Geometry

The cell geometry, that is, the shape of the electrolysis vessel and the positioning and geometric shape of the electrodes, are important in synthetic work. It is very desirable that there be uniformity in the distribution of the electric field between anode and cathode. Uniformity achieves equipotential electrode surfaces, a very desirable, and often necessary, condition for efficient operation and for avoidance of possible overreductions or overoxidations. Equipotential surfaces are best obtained with concentric cylindrical electrodes and with planar parallel electrodes of equal dimensions. Such cells are shown schematically in Fig. 1.4.

In this connection it is appropriate to consider also the technique of measuring the electrode potential. A most practical way is the *Luggin capillary* technique. The Luggin capillary is a convenient means of placing the tip of the reference electrode as close as possible to the surface of the working electrode without causing a significant disturbance in the electric field between anode and cathode. The device used is a slender tube tapering to a narrow tip a few millimeters in diameter; it can be made of glass or of any inert, nonconducting plastic material. The reference electrode, such as a commercial calomel electrode, is positioned in the upper, wider part of the tube. The tube is filled with a conducting solution (salt bridge) or, preferably, with a portion of the solution being electrolyzed. The Luggin capillary is used not only for purposes of precise electrode potential measurements but also for convenience. In many cases the ordinary calomel electrode can be used without the Luggin capillary if the geometry of the cell accommodates it. In most synthetic work consistency in the measurement procedures is more important than very accurate potential measurements that require elaborate measurement techniques, since all electrode potentials are *relative* quantities. *Absolute* electrode potentials are neither experimentally measurable nor necessary for synthetic work.

Once a definite set of electrolysis conditions has been established in a given cell, the current density alone often suffices as a fair criterion for the proper course of an electrolysis. However, the use of a suitable reference electrode to monitor the working electrode potential is always highly recommended, especially during all experimental research work. Figure 1.4 illustrates the use of the Luggin capillary technique. The tip of the capillary is preferably cut at about a 45° angle, and it almost touches the surface of the working electrode. It is important that the tube be free of gas bubbles and that its tip be away from the edges of the electrode. If it is difficult to attain equipotential surfaces, it would be preferable to place the tip of the Luggin in the region of maximum current density to avoid overreductions of overoxidations. Reference electrodes and potential measurements are amply described in the literature and in all electroanalytical textbooks.

1.5   Some Operational Cell Nuisances

The most convenient electrolytic cells are the undivided, or one-compartment, cells. Divided cells are sometimes quite troublesome, especially if the electrolysis medium is an emulsion or a suspension of fine solid particles. All types of diaphragms, of course, add to the overall electrical resistance of the electrolytic system. Sometimes electroosmotic forces cause liquid flow from one compartment to the other, especially if the compositions of the anolyte and catholyte are very different and if the applied cell voltage is more than 15 to 20 V. There is no general solution to this experimental nuisance. Changing the type of diaphragm may alleviate the problem.

Measurement and good control of the electrode potential in heterogeneous media is difficult and of dubious meaning. However, apparent (working) potentials can be established by fixing all other experimental conditions (eg, degree of agitation, cell geometry, position of reference electrode, temperature).

In the course of an electrolysis, the cell voltage sometimes increases substantially, that is, it is sometimes necessary to impose more voltage from the external source in order to maintain the potential of the working electrode at the desired value. This usually happens when poorly conductive metallic oxides, metallic salts, or any other nonconductive films form on the electrode surface. The simplest way to find out which of the electrodes in the cell has built up excess electrical resistance at its surface is to measure the potential of each electrode versus the reference electrode. As a hypothetical case, if the initial cell voltage for a current of 2 A is 2 V and after 2 hr of electrolysis the voltage rises to 10 V (the current may or may not be the same), at least one of the electrodes, usually the anode, has built up excess surface resistance. The potential of that electrode, which might have been +0.6 V initially, may now be found to be +7 V or more. There is no general solution to such a problem. Very often polarity reversals can correct the situation. Chemical or mechanical (scrubbing) cleaning of the electrode can also restore it to its previous condition. The electrolysis, of course, still might be carried out even at the higher cell voltage, provided that the power source has the capacity to keep the potential of the working electrode at the desired value.

It is sometimes necessary to determine which of the electrodes *limits* the current, that is, whether it is the cathodic or the anodic reaction that governs the overall cell reaction rate. A most convenient way to determine this is to lift an electrode only partially from the solution, without disturbing anything else, and observe the change in current. The electrode that causes the decrease in the current when thus lifted is the one that limits the current. Usually, and normally, it is the electrode at which the desired electroorganic reaction occurs. If this is not the case, the counter electrode needs to be larger in area or the concentration of the anodic depolarizer must be increased. The synthesis can be carried out in either case, but it may be unnecessarily slow if its rate is limited by the auxiliary electrode reaction.

1.6   Selection of the Electrode Potential

The meaning of electrode potential and its role in electrochemical synthesis are discussed in some detail in the Appendix. Here we emphasize only that its role in electroorganic synthesis is truly fundamental, since the rate of an electrochemical reaction and the nature of the products, that is, the feasibility of an electrochemical reaction, are determined primarily by the electrode potential. From a purely physical and experimental point of view, the electrode potential is just an *electrical potential*, in a sense analogous to other kinds of potentials, namely, chemical, gravitational, or thermal potentials. From an operational viewpoint in electrochemical synthesis, it can be conceived of as a variable in about the same way as is the wavelength in photochemical reactions or the optimum temperature range in chemical reactions, or as the power of an oxidizing or reducing agent in the usual redox reactions. It is best selected by polarographic or voltammetric methods, as described in a subsequent section. When the electrode potential is *fixed*, as in constant potential electrolysis, its effect is to modify the activation energy barrier so as to affect the reaction rate (electron transfer step). Its effect then is similar in essence to that of a conventional catalyst, inasmuch as catalysts affect the activation energy for a chemical reaction. Thus we may look at the effects of electrode potentials any way the study and its objective are best suited, although the physical reality is one, namely, an electrical potential difference between two phases. Because the potential difference is between two phases, it cannot be experimentally measured in an absolute sense but it can be easily and accurately measured on a relative basis against any convenient reference potential, that is, any chosen reference electrode.

1.7 Divided and Undivided Cells

All cell reactions must involve *at least* two opposite primary reactions: an electronation, or reduction, and a deelectronation, or oxidation. If these reactions do not interfere in any detrimental way (sometimes the interference may be beneficial or even necessary), a one-compartment, or *undivided*, cell can be used for the electrolysis. Otherwise, a two-compartment, or *divided*, cell is required for best results. Consider a hypothetical electroreduction in the two cells shown in Fig. 1.5. Substance A can be reduced at the cathode, but it can just as well be oxidized at the anode. The reduction product B may also be oxidized, directly at the anode or indirectly by some anodically formed substance; this oxidation might give A again or some other product. Under such conditions it would be advisable to employ the divided cell (Fig. 1.5*b*). The diaphragm, represented by the dotted line in Fig. 1.5*b*, prevents both A and B from reaching the anode, and thus only the desired reduction of A to B takes place. The best way to find out the type of cell to be used is by direct experiment.

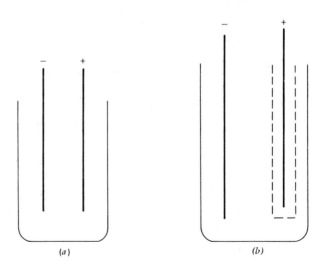

A + e → B   Desired reaction

Figure 1.5   (*a*) Undivided and (*b*) divided cells, showing the need for a diaphragm for reaction A + e → B.

In order for an electric current to pass through the cell, the diaphragm must permit ions, at least of one kind, to pass through it while allowing only minimal solution interdiffusion. The solution in the compartment containing the cathode is called the *catholyte*, while that in the compartment containing the anode is called the *anolyte*. Controlled interdiffusion of anolyte or catholyte is sometimes harmless, or even necessary, for large-scale works. Having thus excluded the possibility of A or B reacting at the counter electrode, we must make it possible for some other electrodic reaction to take place at this electrode; otherwise no cell reaction would be possible at all. The solvent itself, or an ion of the electrolyte, or any other appropriate substance (depolarizer) may fulfill this purpose. This applies also to one-compartment cells. In divided cells, however, the anolyte and catholyte composition may be different not only chemically but even physically. For laboratory cells the diaphragm and the cell design present no serious problems. It is with industrial production cells that the need for a diaphragm is likely to make the electrolytic process unattractive. Every effort should be made to avoid elaborate cell designs and diaphragms when large-scale production is the goal.

This does not mean that diaphragm cells are not practical. The chlor-alkali cells are mostly divided cells, and the largest present electroorganic process, Monsanto's adiponitrile process, employs divided cells. The simple one-compartment cell, however, is always more desirable for large-scale synthesis from both a technical and an economic aspect. For laboratory cells diaphragms are made from various materials: fritted glass, ceramics, various natural and synthetic membranes, ion exchange membranes (permselective membranes), or any suitable cloth of the proper texture. Asbestos has been and still is used as a diaphragm material for large-scale processes and also in the laboratory.

In cases where gases that may form explosive mixtures (eg, H_2, O_2, Cl_2) are produced during the electrolysis, provisions must be made to minimize their mixing. Divided cells can be constructed so as to keep such gases separate. In the case of one-compartment cells an inert gas can act as a diluent by being constantly blown over the cell and thus suppressing the possibility of explosion. A vigorous stream of air, as it passes into the hood where the electrolysis is performed, can normally be effective in preventing the formation of dangerous explosive mixtures of gases.

1.8 Preparation of the Solution for Electrolysis. Solvent and Supporting Electrolyte

All electroorganic syntheses are liquid phase reactions. The organic materials must be in solution in order to exchange electrons with the electrodes. Because most organic substances are not soluble in water, the problem of finding suitable solvents of high dielectric constant can be difficult. Organic polar solvents, such as acetonitrile, dimethylformamide, methanol, sulfolane, and aqueous-organic mixtures, are used for most electroorganic syntheses (see the Appendix). In many cases, and especially when radical cations or anions are formed as primary electrode products, the solvent may be a determining factor in the course of the reaction. The solvent medium must *above all* be able to conduct an electric current. Hence, it must contain at least one component of high dielectric constant that can dissolve and also ionize sufficient amounts of a salt, an acid, or a base, in order for the medium as a whole to be conductive. These ionizable substances are called *supporting electrolytes*, or simply *electrolytes*. Solvents of lower than about 10 dielectric constant may dissolve salts, acids, or bases, but the conductivity of the medium may still be impractically low because the ions may form *ion pairs,* or clusters, which are in effect neutral species, incapable of migrating in an electric field. The solvent should dissolve at least a portion of the organic reactant or reactants. If the organic substance is a liquid, even though it may be insoluble in the medium used, an electrolysis can still be performed by preparing suspensions or emulsions. In such cases vigorous stirring is required. Sometimes an insoluble organic solid material can be dissolved in an inert electrochemically nonpolar solvent, and the resulting solution can then be dispersed in a polar solvent, such as water, for the electrolysis to be carried out.

The solvent, or a component of a mixed solvent, may become involved in the cell reaction intentionally or unintentionally. Many electrosyntheses are effected only through solvent participation in the electrodic reaction. In the process of selecting solvents the literature can be consulted. Various workers have published their findings regarding the *electrochemical domain* of solvents, that is, the potential range within which a solvent is electrochemically inert with a given set of electrodes and electrolysis conditions. Table 1.2 contains data regarding the accessible potentials of commonly used organic solvents. It is, however, always better to determine the practical potential range directly by experiment with the materials at hand under the conditions to be used in the laboratory.

When the ions of the electrolyte function only as carriers of electricity through the electrolysis medium, they are said to be *inert*. However, it should be well understood that *there is no inert electrolyte* in a strict electrochemical sense. The formation of the electrical double layer and its properties as they affect the course of the electrodic reaction is greatly determined by the number and kind of ions of the electrolyte.

Table 1.2   Potential Range (Electrical Domain) of Organic Solvents. Platinum Electrodes, Saturated Calomel Electrode as Reference

Solvent	Electrolyte	Cathodic Limit	Anodic Limit
Acetic acid	NaOAC	−1.0	+2.0
Acetone	$(n\text{Bu})_4\text{NClO}_4$	−1.0	+1.6
Acetonitrile	LiClO_4	−3.0	+2.5
Acetonitrile	$(\text{Et})_4\text{NBF}_4$	−1.8	+3.2
Dimethylformamide	$(n\text{Bu})_4\text{NClO}_4$	−2.8	+1.6
Dimethylsulfoxide	LiClO_4	−3.4	+1.3
Hexamethylphosphoramide	LiClO_4	−3,3	+1.0
Methanol	LiClO_4	−1.0	+1.3
Methanol	KOH	−1.0	+0.6
Methylene chloride	$(n\text{Bu})_4\text{NClO}_4$	−1.7	+1.8
Nitrobenzene	$(n\text{Pr})_4\text{NClO}_4$	−0.7	+1.6
Nitromethane	$\text{Mg}_2(\text{ClO}_4)_2$	−2.6	+2.2
Propylene carbonate	$(\text{Et})_4\text{NClO}_4$	−1.9	+1.7
Pyridine	$(\text{Et})_4\text{NClO}_4$	−2.2	+3.3
Sulfolane	$(\text{Et})_4\text{NClO}_4$	−2.2	+3.0
Tetrahydrofuran	LiClO_4	−3.2	+1.6

The supporting electrolyte's primary function is to carry the current through the solution. (This is the ionic current. The current in the electrodes and in the external circuit is called the electronic current.) Often one of the ions of the electrolyte may also perform the role of a *depolarizer.* A depolarizer is a substance that is preferentially oxidized or reduced at the auxiliary, or counter, electrode so that undesirable reactions at this electrode are minimized. For example, large amounts of hydroxyl ion may act as a depolarizer at the anode, thus suppressing an undesired oxidation of another species (say, of chloride ion). Incidentally, a very good anodic depolarizer is hydrated hydrazine, which, upon anodic oxidation, gives nitrogen gas and protons. The proper depolarizer can reduce the cell voltage and also avoid the necessity for a divided cell.

In organic aprotic solvents the salts most often used as electrolytes are LiClO_4, LiCl, LiBF_4, NaClO_4, $\text{R}_4\text{N}^+\text{BF}_4^-$, $\text{R}_4\text{N}^+\text{X}^-$ (X = halogen), $\text{R}_4\text{N}^+\text{OH}^-$, $\text{R}_4\text{N}^+\text{ClO}_4^-$, and various sulfonates. Perchlorate salts should be handled with great caution, since they pose grave explosion hazards when in contact with organic substances. In aqueous and aqueous-organic media the choice of supporting electrolytes is, of course, greater. Alkali salts, and acids and bases, both inorganic and organic, can be used in aqueous or aqueous-organic media. In methanolic or ethanolic media alkali hydroxides are very good electrolytes. In many electrochemical preparations the role of the supporting electrolyte is fundamental. Substitutions, additions, and indirect oxidations or reductions are possible only through the participation of the ions of the electrolyte, as we see in subsequent sections.

1.9   Temperature Effects

As with all chemical reactions the rates of electrochemical reactions depend on temperature; they normally increase as the temperature increases. The temperature, however, may have some special effects on electrochemical reactions, since the primary reactions occur within the region of the electrical double layer or at the surface of the electrode. Changes in temperature affect adsorption rates, adsorption equilibria, and diffusion rates of reactants and products. Such changes, in turn, influence the course of the overall reaction in various and often unpredictable ways.

Temperature also affects the solubility of the materials used and the conductivity of the electrolytic solution.

1.10   Agitation Effects

It seems obvious that agitation of the electrolysis medium is a key factor, as it is with all heterogeneous liquid phase reactions. The rate of the electrochemical reaction depends on mass transfer, and hence agitation. The nature of the final products may also depend to various degrees on agitation.

Generally, temperature and agitation effects cannot be predicted easily. Only actual preliminary experiments usually point out the proper experimental conditions for the desired objective. Sometimes selectivities are better in two-phase rather than one-phase media. One of the phases extracts the product of the reaction and hinders its further reduction or oxidation at the electrode. In such cases both temperature and agitation may greatly influence the nature of the ultimate products.

1.11   The General Electrochemical Reaction from a Physical Perspective

It would be helpful, and proper, at this juncture to try to view the electrochemical phenomenon from a *physical* perspective. Although all electrochemical cells must contain an anode and a cathode, the primary electrochemical events at each electrode can be considered separately for the purpose of discussion. Let us refer to Fig. 1.1, and consider the processes taking place at one of the electrodes, for example, the cathode. The electrolysis medium contains the ions of the supporting electrolyte, designated as + and −, the organic substrate R to be reduced at the cathode, and species X to be oxidized at the anode. As soon as the power switch is turned on and a dc voltage is applied across the anode and the cathode, the voltmeter will show the magnitude of the applied voltage, while the ammeter will show a current to flow. This current is initially large but it falls rapidly to some steady value, depending on the applied voltage and the composition of the medium. When it is stated that a given voltage has been imposed on an electrolysis cell, it is meant that an electric field of a given magnitude has appeared across the cell. As a result of this field electrically charged species, that is, ions, present in the medium, migrate in opposite directions, negative species to the anode and positive species to the cathode. This directed, field-induced, movement of charged species is called *migration*, and it constitutes the means of current flow in the solution. The current in the solution part of the circuit is called migration or ionic current to distinguish it from the electronic current, which is the flow of electrons in the wires of the external circuit. This movement of charges, ions in the solution and electrons in the wires, would very rapidly cease unless some chemical transformations were taking place at the electrodes. The instantaneous current, or capacitance current, as first shown by the ammeter, decays as soon as the electrical double layers (see the Appendix) at each electrode are built up. In order for a continuous current to flow, substance R must be electronated, while substance X must be deelectronated; to use the more conventional terms, these substances must be reduced and oxidized, respectively. Once these perfectly synchronous events take place, all kinds of chemical reactions are possible. These chemical reactions, which may be followed by further electronations or deelectronations, constitute the practical essence of electroorganic synthesis. The current due to these chemical reactions is referred to as faradaic current to distinguish it from the instantaneous capacitance, or nonfaradaic, current referred to above. (In batteries spontaneous chemical reactions cause the current, whereas in electrolysis the chemical reactions are induced by the imposed voltage, consuming electrical energy.)

Let us now assume that the applied voltage is sufficient so that the organic molecule R can be electronated:

$$R + e \longrightarrow R^{\overline{\cdot}}$$

In order for this molecule to accept an electron from the cathode, it must reach the cathode surface or it must be within a given distance from the surface. It is generally believed, on the basis of experimental observations, that most organic molecules exchange electrons with electrodes from the adsorbed state, that is, the molecule must first be adsorbed on the electrode surface, physically or chemically, in order for the electron transfer act to be possible. In some cases, however, this prerequisite may not exist.

Regardless of how the electron transfer occurs, the organic species (for direct reduction or oxidation) must be in the immediate vicinity of the electrode. If the species carries a charge opposite to that of the electrode, it may reach the electrode by migration, as pointed out above; when the species is uncharged, *diffusion* and *convection* processes must bring it to the electrode. With charged species, of course, all three processes may be operative. Diffusion forces are set up when concentration gradients are established within the electrolysis medium. Concentration gradients are always created at the electrode region as a result of the consumption of the *electroactive* species by the electrodic reaction. Unreacted electroactive species then naturally diffuse from the more concentrated regions to the less concentrated region in the vicinity of the electrode at a rate that is proportional to the concentration gradient. (Diffusion rates increase ~2%/1°C.) If other processes of mass transfer are negligible, the current will be dependent on the rate of diffusion of the electroactive species, and eventually it will become *diffusion limited* in spite of an increase in the applied voltage. While diffusion of the unreacted species occurs from the bulk solution towards the electrode surface, diffusion of the reacted species from the electrode towards the solution also occurs, since the reacted species is more abundant at the electrode surface than away from it. For most practical synthetic electrolyses the transport of the electroactive species, or mass transport, takes place mainly by convectional forces. Convection is achieved by mechanical agitation of the liquid medium and also spontaneously as a result of temperature and density gradient effects. There is always, however, a rather immobile region, the so-called *diffusion region*, in the immediate vicinity of the electrode. With sufficient agitation the *thickness* of the diffusion region layer can be minimized. The diffusion layer thickness is usually of the order of 0.003 to 0.1 mm.

It seems obvious from these considerations that the current, or the rate of an electrodic reaction, can never be greater than the rate of transport of the electroactive species to the electrode, and will always be finite, since the rates of physical processes for mass transport are finite. Although agitation of the electrolysis medium has an effect on the current, it may also have various ef-

fects on the overall course of the electrochemical reaction, since the concentrations of reactants and intermediate products at the vicinity of the electrode are
affected by the degree of agitation. Consider, as an example, only the apparently simple pH effects. Unless sufficient stirring is provided to the electrolysis
medium, changes in pH at the electrode surface in aqueous or aqueous-
organic media might be very large, even in buffered media. Thus electrochemical reactions sensitive to pH may follow a different course, dependent on the
degree of agitation, all other electrochemical conditions being constant.

When the electroactive species reaches the electrode, it finds itself in a
physicochemical environment very different from that in the bulk solution.
Several things may happen in the new environment. The species may become
adsorbed on the electrode, or it may just stay near it at some preferred geomet-
rical orientation as demanded by the electrostatic field forces and other factors at the electrified electrode–solution interface. As with all types of chemical reactions, for any reaction to occur some activation of the reacting species
is required in order to achieve the *transition state* where the electron transfer
step will occur. Both the electron (viewed as a reactant) on the electrode side
and the chemical species on the solution side must acquire the needed activation energy. The electric field (and also the prevailing thermal energy) supplies
this activation energy. Although the temperatures used have no appreciable
effect on the electrons in the metal, they do affect the organic molecule and the
system as a whole, so that the combined effects of electric field energy and
thermal energy create the energy levels in the reacting system which are for the
electron transfer to become possible. Present theories view electron transfer
reactions as adiabatic processes, where the electron transfer is a radiationless
process. In such processes the electronic transitions occur within 10^{-16} sec,
while other molecular events occur within 10^{-11} sec. The electron is transferred
between electronic levels of equal energy in the metal electrode and the reactant species. When it is stated that the rate of an electron transfer process, or
the current (more precisely the net current) at a given electrode potential (and
temperature), attains a certain value, it is implied that the imposition of the
electrical potential, that is, an overpotential, has created a certain amount of
equal electronic energy levels on both sides of the electrode–solution interface
(filled on one side and empty on the other). Under these conditions electrons
can *tunnel* from the electrode to the solution and from the solution to the electrode. The electron tunnels in both directions. However, for a *net* current or
net reaction to be possible, the electron transfer must be faster in one or the
other direction. Otherwise the system remains in dynamic equilibrium, or appears to be so (rest potential). Whether the phenomenon is viewed this way or

in the way of an activation energy hill to be surpassed by the reactants does not really matter, as long as the practical synthetic significance of the electrode potential (overpotential) is understood and appreciated: namely, a net chemical transformation can be possible electrochemically only in the presence of an overpotential. (See also the Appendix, Some Fundamental Electrochemical Concepts and Principles.)

It might be beneficial to form a mental picture of the physical system and the processes described above. The molecules of the organic substrate move (thermally) at random in the bulk solution. The convectional forces that cause them to move bring some of them to the diffusion layer region through which the molecules move towards the electrode by virtue of diffusional rather than convectional forces. The diffusion layer is believed to be quite rigid relative to the rest of the solution. The diffusion layer results from the consumption of the reacting species by the electrodic reaction at the electrode surface. Upon entering the electrical double layer region, where the electric field may be enormous, the organic species suffers an electric *shock*, to use a metaphor. Depending on its nature the molecule tries to recover from this shock by exchanging electrons with the electrode. In this process of electron exchange, however, the molecule usually enters into various chemical reactions, and the electrochemical act ends.

1.12   Preliminary Investigations into the Feasibility of an Electroorganic Synthesis

First, we must clarify the meaning of this section's title. Chemical and electrochemical syntheses, in general, may involve: (1) simple and straightforward reduction or oxidation of functional groups; and (2) all other types of reactions with their various complexities. Electrochemical reactions, without exception, involve at least one reduction or one oxidation at an electrode. The reduction or oxidation may involve: (1) the organic substrate; (2) one or more components of the solvent medium, including the solvent itself; or (3) all of the above. The electrochemical method of synthesis inherently affords easier predictability than does a chemical or a catalytic method. This is because the possibility of the primary reactions, namely, reductions or oxidations, can usually be predicted from simple polarographic and voltammetric observations. There exists a vast literature on this subject (see the General Bibliography).

Of course, the best way to determine the feasibility of any synthesis is to try it on a small laboratory scale. The idea is not so much to try and see what happens, but rather to see how best to try it. It is herein that predictability has meaning and practical value.

The methods of polarography and voltammetry are the most convenient and valuable for predicting the feasibility of most direct electrosyntheses. These methods may also provide clues to the most favorable conditions for an electrosynthesis. In a sense these methods are microscale electrosynthetic techniques in which reactions are not carried out to completion but only for very brief periods so as to establish a *current–potential* relationship character-, istic of a certain electrochemical reaction. These methods are specialty fields in their own right and can be effectively used only by experienced polarographers. A brief qualitative discussion of obtaining current–potential curves by these methods is included here. At the same time the reader is strongly urged to become familiar with these methods and their peripheral modifications. A polarograph and a textbook on polarography are indispensable tools in a modern electroorganic laboratory.

For convenience we may categorize all electrooganic reactions as (1) direct or (2) indirect. In symbols,

$$\begin{cases} R + e \longrightarrow R' \longrightarrow \text{products} \\ R - e \longrightarrow R'' \longrightarrow \text{products} \end{cases} \qquad \text{direct} \quad (1)$$

$$\begin{cases} X \pm e \longrightarrow Y \\ Y + R \longrightarrow RY \\ \text{also } Y + R \longrightarrow R' + X \end{cases} \qquad \text{indirect} \quad (2)$$

The electrode reactions represented by eq. (1) are direct in the sense that the organic substrate, R, exchanges electrons directly with the electrode. A substance that can exchange electrons directly with the electrode is said to be electroactive. Reaction (2) is indirect with regard to the organic substrate R, but direct with regard to X becoming Y. Polarography and voltammetry are usually unable to show the possibility of indirect electrolytic reactions. They can, however, show the feasibility of direct reactions, provided such reactions occur at a potential lower than the one at which the solvent or the electrolyte would be oxidized or reduced at the electrode at which the organic substance of interest is studied.

A substance, to be electroactive, must contain one or more *electrophores*. An electrophore is a group of atoms in a molecule capable of accepting or releasing electrons at an electrode at a given potential. For example, the nitro group in nitrobenzene or the amino group in aniline are electrophores. So is the entire anthracene molecule when directly oxidized or reduced at an electrode, and the carbon-chlorine bond in pentachloropyridine when cathodically reduced to a carbon-hydrogen bond. The term electrophore is best used by analogy with the term chromophore in photochemistry.

In conventional chemical syntheses we normally think of *single* reactions:

$$A + C \longrightarrow B + D \tag{3}$$

In electrolytic cell reactions we must always think of *paired* reactions, one taking place at the anode and the other at the cathode, and *neither* ever being possible without the other:

$$A + e \longrightarrow B \qquad \text{cathodic} \tag{4}$$

$$C - e \longrightarrow D \qquad \text{anodic} \tag{5}$$

The cell reaction is the sum of reactions (4) and (5):

$$A + C \longrightarrow B + D \tag{6}$$

Polarography and voltammetry, as we see, give information about either the anodic or the cathodic reactions but not about the overall cell reaction as represented by eq. (6). These methods show the presence or absence of electrophores in a molecule and the relative ease of their reduction or oxidation. In general, electrophores are oxidized or reduced at specific potentials. The *half-wave* potentials, $E_{1/2}$, obtained from polarographic current-potential curves have both practical and theoretical value. For reversible electrode reactions the $E_{1/2}$'s are almost identical with *standard redox* potentials. For slow or irreversible reactions the $E_{1/2}$ has no thermodynamic meaning but the entire polarogram is a *picture* of the kinetics of the electrochemical reaction. A vast literature is available on this subject.[3] When considering electroactivity this literature can be profitably consulted. It is, however, very easy, and always advisable, to determine the electroactivity of a substance in the laboratory, and under the intended experimental conditions. It should be borne in mind that electroactivity often depends on all three of these factors: (1) the nature of the organic substance; (2) the electrode; and (3) the electrolysis medium. *Electroactivity is a property of a given electrochemical system in its entirety rather than an independent molecular property.* Briefly, the experimental procedure for obtaining a polarogram, or voltammogram, is as follows:

Consider the three-electrode cell in Fig. 1.6a. In this cell the *working* electrode, W, is a microelectrode of a few square millimeters total surface area. In polarography this is the *dropping mercury* electrode (DME) and in voltamme-

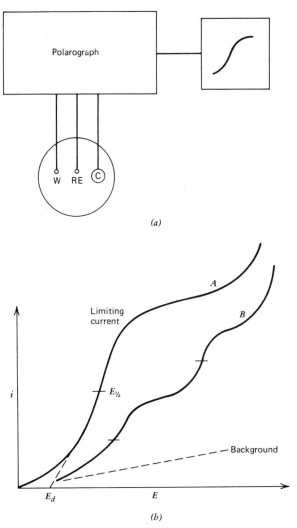

Figure 1.6 (a) Schematic of polarographic cell. (b) polarograms (voltamograms). W: working electrode. C: auxiliary (counter) electrode. RE: reference electrode.

try any electrode, such as a tiny piece of a metallic wire. The working electrode must be made of the same material as the large-scale electrolysis electrode, if useful correlations are to be made.

The *auxiliary*, or *counter*, electrode, C, is much larger than W. This large difference in size is necessary in order to obtain polarograms or voltammograms. Because of its much smaller size, the working microelectrode becomes *concentration polarized;* this enables the development of a polarographic or voltammetric *wave,* as illustrated in Fig. 1.6. The reaction to be studied is that occurring at the microelectrode. The third electrode, RE, is a reference electrode, such as a saturated calomel electrode, against which the potential of the microelectrode is measured. The potentiostat impresses a dc voltage across the two electrodes, W and C, one of which is the anode and the other the cathode. Initially, as shown in Fig. 1.6*b*, only a small current is observed. As the applied voltage on the cell increases, as a result of which the potential of the microelectrode also increases, the current rises slowly until a potential is reached beyond which the current rises steeply, as shown by the rising portion of the curve. The potential, shown as E_d, is the *decomposition potential* of the substance being reduced or oxidized at the microelectrode (this potential must not be confused with the *cell decomposition* potential; see the Appendix). If the solution is not stirred, a *limiting current* is eventually established, as shown by the upper plateau region of the curve. In this region, the microelectrode is *concentration polarized.* In this potential region the rate of the electrodic reaction is limited by the rate of mass transfer processes, not by the activation energy for the electron transfer reaction. The current is no longer a function of electrode potential but rather of another potential, namely, the *diffusion potential.* Because the rate of diffusion of the electroactive species to the electrode surface is constant, the rate (current) of the electrochemical reaction is also constant and proportional to the rate of diffusion of the electroactive substance. The rate of diffusion is proportional to the bulk concentration of the electroactive substance. Because of this proportionality, limiting currents are of valuable analytical importance. For the best polarograms or voltammograms (waves) to be developed, concentrations of the electroactive species should be in the range of 10^{-2} to 10^{-4} M. The limiting, or diffusion, current develops best when almost all the *migration* current is suppressed. This is done by using a large amount of supporting electrolyte relative to the concentration of the electroactive substance. The point in the middle of the rising part of the curve is the *half-wave potential,* $E_{1/2}$, of the particular reaction under the experimental conditions, and it is characteristic of the molecule reduced or oxidized as an entity and also of a particular electrophore. It may or may not have thermodynamic significance ($\Delta G = nFE^0 \sim nFE_{1/2}$ for reversible processes only; for irreversible processes $E_{1/2}$ can be related to the activation energy for the process). On the basis of differences in the $E_{1/2}$'s of various substances and electrophores in a molecule, selective electrolytic syntheses may be effected that are not possible by conventional chemical or catalytic methods. In practice selective electrolyses can be achieved when the $E_{1/2}$'s differ by at least 0.1 V for laboratory-scale syntheses and by more than 0.2 to 0.3 V for large-scale syntheses.

If we assume that curve A in Fig 1.6b represents a cathodic reaction:

$$R + e \longrightarrow R^{\bar{\cdot}} \xrightarrow{\text{H+}} RH$$

and if we wish to perform an electrolysis for the preparation of RH, it would be obvious from the polarogram that a potential in the region of the rising portion of the curve should be used, preferably one corresponding to the upper shoulder of the wave. Curve B, with two waves, usually indicates one of the following possibilities: (1) a stepwise electron transfer reaction; (2) two different electrophores; (3) two chemically similar electrophores on a molecule, that are of different electroactivity; or (4) two different substances. Symbolically,

$$R + e \longrightarrow R^{\bar{\cdot}} \xrightarrow{e} R^{=} \tag{1}$$

$$\begin{cases} XAY + e \longrightarrow XA\dot{\bar{Y}} \longrightarrow \text{product} \\ XAY + e \longrightarrow \dot{\bar{X}}AY \longrightarrow \text{product} \end{cases} \tag{2}$$

$$XAX + e \longrightarrow \dot{\bar{X}}AX \xrightarrow{e} \dot{\bar{X}}A\dot{\bar{X}} \tag{3}$$

$$\begin{cases} Y + e \longrightarrow Y^{\bar{\cdot}} \\ Z + e \longrightarrow Z^{\bar{\cdot}} \end{cases} \tag{4}$$

If an organic substance does not appear to be polarographically or voltammetrically active, it still might be electrochemically active from a preparative viewpoint. As pointed out above, polarographic waves develop well only when the solvent or the supporting electrolyte remains inactive within the potential range used. If both the medium and the organic substance under study are oxidized or reduced, no wave develops because the needed concentration gradient at the electrode surface cannot be established. Instead of the plateau observed under diffusion controlled reactions, the current continues to rise with the potential until other causes eventually force it to change in one direction or the other. It should be pointed out that limiting currents are not due only to diffusion but also to several other causes. Thus besides diffusion-limited currents there are also *adsorption, catalytic,* and *kinetic* currents. Adsorption limited currents are caused by adsorption of the oxidized or reduced form of the substrate at the electrode surface. Catalytic currents are limited by the catalytic effectiveness of the catalyst, and kinetic currents by the rate of the preceding chemical step (CEC . . . mechanism).

An electroorganic synthesis does not always require that the organic substrate itself be electroactive. The desired reaction may be indirect. Current–potential curves in such cases may represent the electroactivity of the substance X in the reaction scheme:

$$X \pm e \longrightarrow Y$$
$$Y + R \longrightarrow RY$$

Here, for example, X may be a chloride ion, and R an organic compound, such as propylene:

$$2Cl^- \longrightarrow Cl_2 + 2e$$

$$\downarrow \begin{array}{c} CH_2{=}CHCH_3 \\ H_2O \end{array}$$

$$ClCH_2CHOHCH_3$$

The voltammogram will show the needed anode potential for the formation of chlorine, which would react with propylene, as shown above, but not the probability of the following chemical hydrochlorination reaction, although this may sometimes be inferred from the shift of the voltammogram to less anodic values or, in the case of reductions, less cathodic values.

In conclusion, polarographic and voltammetric studies and a consultation of the literature can be of great value in predicting and planning an electroorganic synthesis. The experimenter, nonetheless, should try the desired synthesis even if the preliminary microscale observations appear to be inconclusive. Also, it must be borne in mind that microscale and macroscale happenings do not always agree qualitatively or quantitatively. (This has been well emphasized by P. Zuman.)

One note must be made: If a substance is known to be oxidized or reduced in a given medium by chemical reagents, it is very likely that this substance would be electroactive provided the proper solvent medium and electrode for carrying out the electrolysis can be found. *There are no inherently nonelectroactive substances.* What happens is that the needed electrode potential for the organic substance is beyond the electrochemical domain of the medium and therefore the needed potential for the organic substance cannot be reached experimentally.

References

1   J. O'M. Bockris and A. K. N. Reddy, *Modern Electrochemistry*, Vol. 2, Plenum, New York, 1970.

2   N. L. Weinberg, Ed., *Techniques of Electroorganic Synthesis*, Part I, Wiley-Interscience, New York, 1975.

3   A. J. Bard, *Encyclopedia of Electrochemistry of the Elements* (Organic Section), Vols. XII ff, Marcel Dekker, New York, 1973; L. Meites and P. Zuman, *Handbook Series in Organic Electrochemistry*, Vols. I ff, CRC Press, Boca Raton, Florida, 1978; N. L. Weinberg, Ed., *Technique of Electroorganic Synthesis*, Part II, Wiley-Interscience, New York, 1975.

Following are some references to advanced books and theoretical aspects of electrochemistry.

1   J. O'M. Bockris, *J. Chem. Educ.*, **48**(6), 353 (1971) (about concept of overpotential).

2   J. O'M. Bockris and A. K. N. Reddy, *Modern Electrochemistry*, Vols. 1 and 2, Plenum, New York, 1970.

3   B. E. Conway, *Theory and Principles of Electrode Processes*, Ronald, New York, 1965.

4   P. Delahay, *New Instrumental Methods in Electrochemistry*, Interscience, New York, 1954.

5   P. Delahay and C. Tobias, *Advances in Electrochemistry and Electrochemical Engineering*, Several volumes, continuing, Interscience, New York, 1961.

6   H. Eyring, Ed., *Physical Chemistry, An Advanced Treatise*, Vol. IXA, *Electrochemistry*, Academic, New York, 1970.

7   K. J. Vetter, *Electrochemical Kinetics*, Academic, New York, 1967.

8   P. Zuman, *Prog. Phys. Org. Chem.*, **5**, 161 (1965).

2

A BRIEF SURVEY OF ELECTROORGANIC REACTIONS

ANODIC REACTIONS

2.1 Hydrocarbons

The anodic oxidation of hydrocarbons is almost as easy as their chemical oxidation. Condensed aromatics, such as anthracenes and naphthalenes, and unsaturated aliphatics are oxidized more easily than saturated hydrocarbons. The mode of oxidation can be direct or indirect. Saturated hydrocarbons require anodic potentials of 3 to 3.5 V (vs the usual reference electrodes) to be directly oxidized. Very few solvents can remain inert at such high potentials. This makes direct oxidation of saturated hydrocarbons practically unfeasible. They can, however, undergo indirect anodic oxidations. For example, they may react with electrolytically generated hydroxy radicals or any other anodically produced active species:

$$OH^- \longrightarrow OH^{\cdot} + e$$
$$RH + OH^{\cdot} \longrightarrow {\cdot}RHOH$$
$$Br^- \xrightarrow{-e} Br{\cdot} \xrightarrow{RH} Br\dot{R}H$$
$$\downarrow {\scriptstyle -H^{\cdot}}$$
$$RBr$$

The reactions shown are difficult to control for synthetic work, but the possibility of utilizing them exists. It is generally accepted that reactions of this type occur while the reactants are adsorbed at the anode surface.

It has been shown by Bertram et al[1] that aliphatic hydrocarbons can be directly oxidized in fluorosulfonic acid containing sodium fluorosulfonate (the acid by itself is a poor conductor). Thus it has been possible to obtain α,β-unsaturated ketones from alkanes by electrolysis in fluorosulfonic acid containing various carboxylic acids. Cyclohexane, for example, seems to go through the cyclohexenyl cation to the five carbon ring ketonic product:

$$C_6H_{12} \xrightarrow{-2e}$$

It is not clear how the strong acidic solvent makes direct electron transfer easier than do other media. Protonation of the alkane has been invoked as an explanation, but we would expect such an event to favor reduction rather than oxidation. It appears that the strong acidic medium weakens, somehow, the C—H bond of the alkane and thus makes the electron transfer to the anode easier.

In contrast to saturated aliphatics, unsaturated and aromatic hydrocarbons are oxidized directly, requiring only about 1 V anodic potentials.[2-5] These oxidations involve electrons of π systems and are usually two-electron oxidations:

Cyclic voltammetry indicates that the first electron transfer is, in many cases, reversible. The expulsion of a proton and another electron results in a cation that then reacts with the medium to give the products.

It is most likely that the aromatic molecule is adsorbed at the anode surface in a flat position or in a position such that the π electron cloud has maximum contact with the electrode surface. This would make the electron transfer easiest, and it is under such an arrangement that the electron transfer might be rapid and reversible on a cyclic voltammetric time scale. If the molecule continues to be adsorbed even after the second deelectronation and the deprotonation step, stereochemical substitutions not possible by conventional chemical reactions can be achieved. With unsaturated species additions would be possible that might also be of stereochemical importance. The possibilities of substitutions and additions are many.

2.1.1   Acetoxylations

Schematically, an anodic acetoxylation of an aromatic hydrocarbon can be conceived of in three possible ways[6,7]:

$$\textbf{1} \quad ArH \xrightarrow{-e} ArH^{\ddag} \xrightarrow{OAc^-} \dot{A}rHOAc$$

$$\downarrow{\scriptstyle -H^{\cdot} \mid -e}$$

$$ArOAc$$

$$\textbf{2} \quad ArH \xrightarrow{-e} ArH^{\ddag} \xrightarrow{-e} ArH^{2+}$$

$$\downarrow{\scriptstyle -H^{\cdot} \mid OAc^-}$$

$$ArOAc$$

$$\textbf{3} \quad \text{electrode} \;\; ArH \quad \overset{O}{\underset{O}{\overset{\|}{CCH_3}}} \xrightarrow[-H^{\cdot}]{-e} ArOAc$$

Acetoxylations can take place in media consisting of an organic solvent, such as dimethylformamide, acetonitrile, acetic acid, sulfolane, and mixtures of such solvents, and in the presence of fairly large amounts of an acetate salt, usually sodium acetate. Almost all kinds of anode materials can be used. The anode potential must be positive enough to deelectronate the organic substrate in order for the acetoxylation to be possible, but not so positive (more than +2.0 V) that it causes oxidation of the acetate ion itself. (Oxidation by acetoxy radical $CH_3COO^{\cdot}$ does not seem likely, since such a species is known to be unstable.) Acetoxylations occur before the acetate ion is oxidized. This is evidence that oxidation of the organic substrate occurs to form the cation or cation radical, which then reacts chemically with the nucleophilic acetate species. In the presence of other nucleophiles competitive substitutions or additions would occur. The last expression is intended to show the possibility of a concerted mechanism, that is, the simultaneous electron transfer to the anode and the attack by OAc⁻, one act assisting the other. An interesting stereochemical example of anodic addition is the acetoxylation of stilbene[8]:

$$PhCH{=}CHPh \xrightarrow{-e} Ph\overset{+}{C}H{-}\overset{\cdot}{C}HPh$$

$$2AcO^- \downarrow -e$$

By hydrolysis mono- and dihydroxy derivatives are obtained.

2.1.2 Cyanations

The cyanide ion itself is easily oxidized at the anode (~1.0 V at platinum and graphite). In spite of the easy oxidation of this ion most reported cyanations suggest that this ion acts as a nucleophile (being a potent one) on the anodically formed organic species, since no cyanations occur before oxidation of the organic species occurs.[9,10] The cyanation reaction can therefore be expressed in a manner analogous to that of acetoxylations:

It can also be conceived of as a homolytic substitution:

$$CN^- \xrightarrow{-e} CN^\cdot \xrightarrow{RH} RCN + H^\cdot$$

but available experimental evidence rather negates this possibility.[9] Methanolic solutions with sodium cyanide have been used for cyanations.[11–13] Cyanations of an aromatic hydrocarbon could be visualized at an adsorbed state, perhaps assisted by the negative cyanide ion (although at the positive potentials used, oxidation of the cyanide ion also occurs):

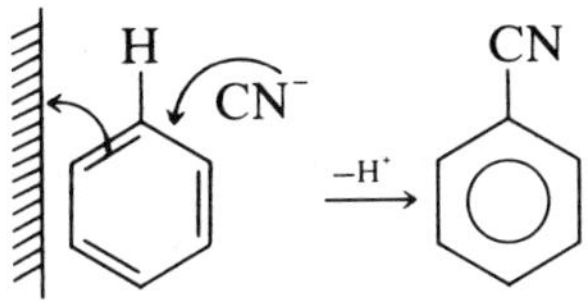

Anodic oxidations of hydrocarbons in the presence of the cyanide ion can be performed in various media, provided the medium itself is a weaker nucleophile than the cyanide ion. Of course, we must also consider the kind of cell in such reactions. A solvent that is suitable in a divided cell may be entirely unsuitable in an undivided one. In the latter cell the solvent itself may be cathodically converted to a nucleophilic species and may thus be competitive with the cyanide. Cyanation can take place at both the aromatic nucleus and the side chain in a methanol-sodium cyanide medium.[14]

2.1.3 Methoxylations

Methanol can be anodically oxidized, as can other alcohols in general. With suitable organic substrates methoxylations can be accomplished in a manner analogous to acetoxylations and cyanations. It is not very clear whether the reaction mechanisms are radical or polar. Some experimental results favor a radical while others favor a polar mechanism. At the anode a methanol molecule deelectronates to form a free radical and a proton:

$$CH_3OH \xrightarrow{-e} CH_3O^{\cdot} + H^+$$

Methoxide ion, when sodium methoxide-methanol media are used, oxidizes to give the methoxy radical. A methoxylation reaction can then be expressed as

$$CH_3O^{\cdot} + RH \longrightarrow R^{\cdot} + CH_3OH$$

The radical $R^{\cdot}$ is oxidized at the anode to give the cation R^+, which reacts with either methanol or methoxide ion:

$$R^{\cdot} \xrightarrow{-e} R^+ \xrightarrow[-H^{\cdot}]{CH_3OH} ROCH_3$$

Direct deelectronation of a hydrocarbon molecule may also lead to methoxy products:

$$ArH \xrightarrow{-e} ArH^{\ddagger} \xrightarrow[\substack{-e \\ -H^{\cdot}}]{CH_3O^{-}} ArOCH_3$$

In such cases we would expect a base in the medium to be beneficial in that it would abstract the proton from the anodic intermediates. The possibility of two-electron oxidation of methanol may be considered:

$$CH_3OH \xrightarrow[-H^{\cdot}]{-e} CH_3\overset{\cdot}{O} \xrightarrow{-e} \begin{array}{c} H \\ | \\ HC-\overset{+}{O} \\ | \\ H \end{array}$$
$$H_2\overset{+}{C}OH$$

Parker[15] observed that electrolysis of a solution consisting of methanol, 1% H_2SO_4, and anisole, gave *p*-methoxybenzyl alcohol; the following explanation was given as

$$CH_3OH \xrightarrow{-e} CH_3O^{\cdot} \xrightarrow{CH_3OH} HO\overset{\cdot}{C}H_2$$
$$HO\overset{\cdot}{C}H_2 \xrightarrow{-e} HO\overset{+}{C}H_2$$

$$HO\overset{+}{C}H_2 \xrightarrow{anisole} \underset{CH_2OH}{\overset{OCH_3}{\bigcirc}}$$

In this acidic medium formation of methyl sulfate and disulfate is likely to give cation at the anode, such as

$$CH_3OSO_3H \xrightarrow[-H^{\cdot}]{-2e} \overset{+}{C}H_2OSO_3H$$

This can conceivably attack the anisole to give, upon hydrolysis, the product shown previously.

2.1.4 Anthracenes

The anodic oxidation of anthracenes has been studied by Lund[16] in aprotic media. In an acetonitrile/perchlorate/pyridine medium, the salt 9,10-dihydroanthranylpyridinium diperchlorate was formed. This is consistent with a two-electron oxidation and indicates the stability of the formed dication in the acetonitrile/pyridine medium:

The pyridine is a stronger nucleophile than the solvent acetonitrile. If water is present in acetonitrile it adds to the dication to form anthraquinone and bianthrone,[17] as we would expect. Parker reported an interesting stereochemical anodic oxidation of anthracene in an acetonitrile/acetate medium.[18] Chemical oxidation with lead tetraacetate yields equal amounts of the *cis-* and *trans-*diacetate isomers, while anodic oxidation yielded mostly the *trans-*diacetate isomer.

2.1.5   Oxidation of Alkylaromatics

Alkyl substituted aromatic hydrocarbons may lose, upon oxidation, a proton from the alkyl chain to give the chain substituted products. This would be shown as

$$ArCH_3 \xrightarrow{-e} ArCH_3^{+} \xrightarrow[-H^{\cdot}]{-e} Ar\overset{+}{C}H_2$$
$$Ar\overset{+}{C}H_2 + X^{-} \longrightarrow ArCH_2X$$

If the electrolysis medium lacks nucleophilic properties and if the supporting electrolyte is a salt whose anion cannot form bonds with the organic substrate, that is, if salts such as $Bu_4N^+ClO_4^-$ and $Bu_4N^+BF_4^-$ and solvents such as methylene chloride, nitromethane, and sulfolane are used, the possibility of polymeric products is greatly enhanced, as it would be if chemical methods were used to produce the radical cations and free radicals:

$$ArCH_3 \xrightarrow[R_4N \cdot BF_4]{-e} ArCH_3^{+} \xrightarrow[-H^+]{-e} Ar\dot{C}H_2$$

The free radical can dimerize, or it can be further oxidized to give the carbocation, $ArCH_2^+$, and finally various products derivable from such intermediates, depending on concentrations of starting materials, current densities, temperature and agitation, and electrode surface characteristics. With mixtures of nucleophilic solvents complex product mixtures would be expected.

In acetonitrile with small amounts of water the following reaction occurs:

$$ArCH_2^+ + CH_3CN \xrightarrow{H_2O} ArCH_2NHCOCH_3$$

Extensive studies on this type of reaction have been reported by Eberson and Nyberg.[19]

Anodic oxidations of aromatics can be achieved in both acidic and basic aqueous or aqueous-organic media. If oxygenated[20] substances are desired, a source of oxygen donors must be present in the electrolysis medium. Deelectronations are defined as oxidations, but the term oxygenation is preferred if substitutions by, or additions of, oxygen are implied. Aqueous media containing acids whose anions do not usually form bonds with carbon atoms (H_2SO_4 or $HClO_4$) can be used for anodic oxygenations. Catalytic amounts of metallic ions can sometimes be employed for such reactions. Oxidations in 30% H_2SO_4 and 15% NaOH aqueous media have been carried out as emulsions with organic hydrocarbons.[20] Mesitylene and pseudocumene gave their aldehyde derivatives as major products. Tomat[21] has shown the catalytic effect of the ferric-ferrous oxygen system in the electrolytic oxygenation of polymethylated benzenes. The methyl group was thus converted to an aldehyde function by indirect oxidation via OH· radicals:

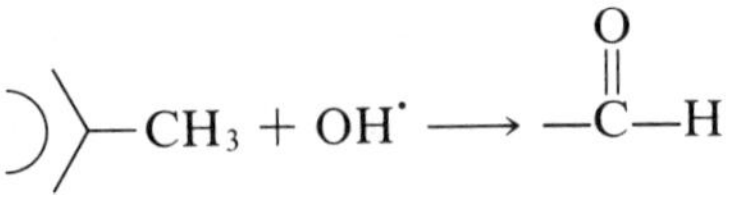

Apparently, the electrolytically regenerative system, $Fe^{2+} + H_2O_2$, was involved in the oxidation. We are also free to speculate about using organic and organometallic catalysts produced *in situ* or *ex situ*. Anodic additions to unsaturated hydrocarbon systems, especially conjugated systems, should be possible. Consider these hypothetical reactions:

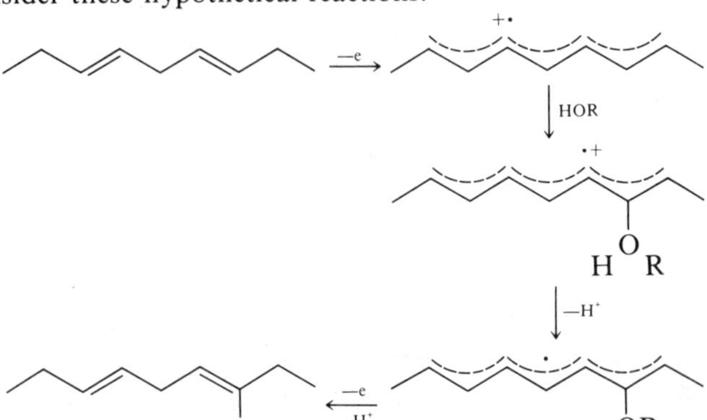

With difunctional compounds, such as glycol, this addition would be expected:

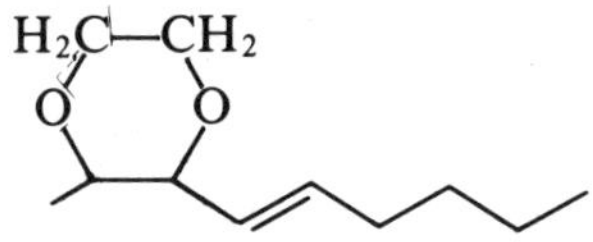

In the presence of water epoxide formation would also be possible. A very interesting study on the anodic addition of ureas and ethylene glycol to conjugated dienes was recently reported by Baltes et al.[22] The anodic reactions were interpreted as involving the primary anodic radical cation:

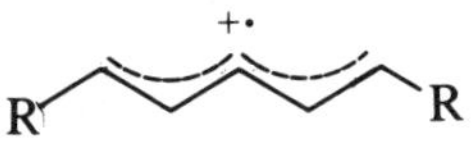

Products such as the following were obtained:

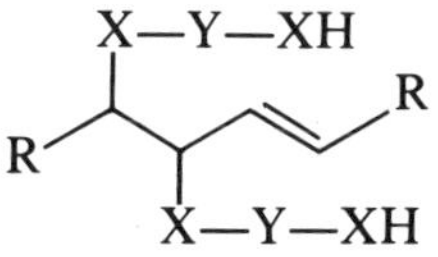

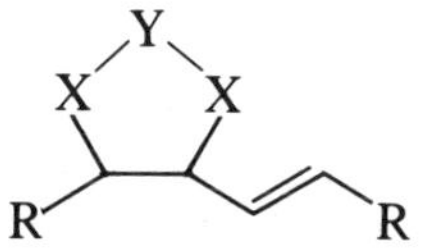

where X is NCH_3, O, NC_6H_5 and Y is CO, $-CH_2CH_2-$.

These types of additions can also be effected by homogeneous chemical reactions. The electrolytic reaction would be advantageous if selective and stereoisomeric additions were possible as a result of orientation and adsorption of reactants at the anode surface.

At this juncture we make a very brief digression to try to see the similarity between photochemical and electrochemical events. Consider, for instance, the anodic oxidation of triphenylmethane[23]:

$$Ar_3CH \rightleftharpoons [Ar_3CH]^{\ddagger} + e$$
$$\Updownarrow$$
$$Ar_3CH^{2+} + e$$
$$\downarrow{\scriptstyle -H^{\cdot}}$$
$$Ar_3C^+$$

The photochemical oxidation is expected to proceed in the presence of oxygen as follows:

$$Ar_3CH + O_2 \xrightarrow{h\nu} [Ar_3CH]^{\ddagger} + O_2^{-}$$
$$\downarrow O_2^{-}$$
$$Ar_3C^{+} + O_2H^{-}$$

Here there is no anode to accept the electron from the organic molecule. Instead, an oxygen molecule accepts the electron. In the electrochemical electron transaction the electric field (the overpotential) provides the needed energy, while in the photochemical case the electromagnetic field ($h\nu$) is called upon to provide the energy. In the former case the events are most likely to be heterogeneous, that is, the reactants are adsorbed and more or less oriented at the electrode–solution interface, whereas in the latter case such a restriction is nonexistent, and the reaction is homogeneous (photochemical reactions may also be advantageously carried out under heterogeneous conditions, however).

As long as we are discussing the oxidation of hydrocarbons, we would be somewhat remiss not to mention the fact that carbon itself, in the form of dispersed coal particles, can be electrolytically converted to CO_2, $RCOO^-$, and R—R species at the anode in an aqueous medium while hydrogen gas is liberated at the cathode.[24]

References

1   J. Bertram, J. P. Coleman, M. Fleischmann, and D. Pletcher, *J. Chem. Soc., Perkin II*, **1973**, 374.

2   J. W. Loveland and G. R. Dimerel, *Anal. Chem.*, **33**, 1196 (1961).

3   W. C. Neikam, G. R. Dimerel, and M. M. Desmond, *J. Electrochem. Soc.*, **111**, 1190 (1964).

4   M. Fleischmann and D. Pletcher, *Tetrahedron Lett.*, **1968**, 6255.

5   H. Lund, *Acta Chem. Scand.*, **11**, 1323 (1957).

6   L. Eberson, *J. Am. Chem. Soc.*, **89**, 4669 (1967).

7   S. D. Ross, M. Finkelstein, and R. C. Petersen, *J. Org. Chem.*, **35**, 781 (1970).

8   F. D. Mango and W. A. Bonner, *J. Org. Chem.*, **29**, 1367 (1964).

9   V. D. Parker and B. E. Burget, *Tetrahedron Lett.*, **1968**, 2415.

10   L. H. Klemm, P. E. Iversen, and H. Lund, *Acta Chem. Scand.*, **B28**, 593 (1974).

11   K. Koyame, T. Susuki, and S. Tsutsumi, *Tetrahedron Lett.*, **1965**, 627.

12 L. Eberson and S. Nilsson, *Discuss. Faraday Soc.,* **45**, 242 (1968).

13 S. Andreades and E. W. Zanow, *J. Am. Chem. Soc.,* **91**, 4181 (1969).

14 K. Yoshida and S. Nagase, *J. Am. Chem. Soc.,* **101**, 4268 (1979).

15 V. D. Parker, *Chem. Ind.* (London), **1968**, 1363.

16 H. Lund, *Acta Chem. Scand.,* **11**, 1323 (1957).

17 V. D. Parker, *Acta Chem. Scand.,* **24**, 2757 (1970).

18 V. D. Parker, *Acta Chem. Scand.,* **24**, 3162 (1970).

19 L. Eberson and K. Nyberg, *Am. Chem. Soc. Symposium,* Chicago, Sept. 13–18, 1970.

20 L. G. Feoktistov, *Electrokhim. Org. Soedin, Tezisy Dokl.,* 1976 (*CA*, **87**, 13442₅, 1977; **86**, 62685ₘ, 1977).

21 R. Tomat, *J. Appl. Electrochem.,* **9**, 301 (1979).

22 H. Baltes, L. Stork, and H. Shäfer, *Liebigs Ann. Chem.,* **1979**, 318.

23 J. E. Kuder, W. W. Limburg, M. Stolka, and S. R. Turner, *J. Org. Chem.,* **44**, 761 (1979).

24 R. W. Coughlin and M. Farooque, *Nature,* **279**, 301 (1979).

2.2 Hydroxy Compounds

Hydroxyl groups are oxidized anodically with relative ease. Primary alcohols may give aldehydes, carboxylic acids, acetals, esters, ethers, and even hydrocarbons as a result of decarboxylation of formed carboxylic acids.[1-4,13] The formation of acetals, for example, can be visualized as

$$RCH_2OH \xrightarrow{-e} [RCH_2OH]^{\ddagger}$$

$$\downarrow{\scriptstyle -H^+ \mid -e}$$

$$\overset{+}{R}CHOH$$

$$\overset{+}{R}CHOH + RCH_2OH \xrightarrow[RCH_2OH]{-H_2O} RCH(OCH_3)_2$$

There is apparently no general mechanism for the anodic oxidations of alcohols. Both the anode material (and surfaces) and the electrolysis medium have large effects on the mechanistic pathway and the ultimate products. In attempting anodic oxidations with alcohols acidic media should be preferred, since in basic media aldehydes undergo various side reactions that may be undesirable. Sulfuric acid media would be quite suitable for many such oxidations.

The oxidation of benzyl alcohol to benzaldehyde and benzoic acid has been formulated as

Benzylic ethers and esters are also formed.

The kinetics of anodic oxidation of alcohols in basic media using various oxide-covered anodes suggest indirect rather than direct oxidations of these substances. The metallic oxide was believed to cause dissociation of adsorbed alcohol by abstracting hydrogen from the molecule.[4] Oxidation of phenylethanols, alkyl alcohols, and benzyl alcohols in alkaline water/t-butyl alcohol media afforded carboxylic acids. In these cases the primary electrodic reaction at the anode is the formation of metallic surface oxides and peroxides that chemically react with the adsorbed organic molecules. In the process the surface oxidants are reduced by the alcohol to be again oxidized anodically and thus (catalytically) repeat the cycle with a new molecule of alcohol. Using nickel anode material the mechanism was proposed as follows:

$$Ni \xrightarrow{H_2O/KOH} Ni(OH)_2 \underset{\text{surface}}{\overset{-e}{\rightleftharpoons}} NiO(OH) + H^+$$
$$\phantom{Ni \xrightarrow{H_2O/KOH}} \text{surface}$$

$$NiO(OH) + RCH_2OH \longrightarrow Ni(OH)_2 + R\dot{C}HOH$$

The surface reaction with the alcohol could be viewed as "dissociative adsorption."

A potentially practical synthesis with electrogenerated halogens was reported recently.[5] Primary alcohols were oxidized to acetals of 2-haloaldehydes. Acidic methanolic media, using anhydrous hydrogen halides, were used (the possibility of forming methyl hypochlorite, an explosive substance, must be borne in mind). It is difficult to explain the apparent inertness of the methanol medium. It would perhaps be reasonable to suggest that methanol is also oxidized to some extent by chlorine but the product is reduced rapidly at the cathode.

Shono et al achieved intramolecular cyclization with olefinic hydroxy compounds[6]:

The deelectronation appears to involve the olefinic bond; it is not clear, however, which function, the olefinic bond or the hydroxyl, is the electroactive center of the molecule.

An intramolecular carbon-oxygen cyclization involving the aromatic hydroxyl is the electrooxidation of tyrosyl compounds[7]:

The following intermolecular coupling reaction of the isoquinoline alkaloid hydroxy compound has been observed[8]:

The role of the hydroxyl function here is not clear.

Anodic oxidation of sterically hindered phenols in the presence of certain nucleophiles has been shown to afford addition products via the phenoxonium ion[9]:

Dihydric phenols are oxidized more easily than monohydric phenols,[10] as is apparent from their oxidation potentials [vs the saturated calomel electrode (SCE)]: phenol, 0.53 V; cathecol, 0.139 V; hydroquinone, 0.018 V. Phenol gives high yields (90%) of hydroquinone under controlled conditions.[11]

Oxidation of alcohols has been achieved by what seems to be indirect catalytic electrolysis.[12] Thus secondary alcohols were oxidized to the corresponding ketones by electrolysis in butyl alcohol or butyl alcohol/hexane mixtures in the presence of potassium iodide. The iodonium ion, I^+, formed electrolytically, was proposed to be the oxidizing agent, and also the catalyst, as shown:

$$I^- \xrightarrow{-2e} I^+$$
$$I^+ + R_2CHOH \longrightarrow R_2CO + I^- + 2H^+$$

Oxidation of primary alcohols at nickel anodes in aqueous sodium hydroxide media was reported to give carboxylic acids with both short and long chain alcohols.[13] Aqueous dispersions of the alcohols or aqueous-organic mixtures in which the alcohols are soluble can be used for the electrolysis.

In certain cases of anodic oxidation of phenolic compounds, deprotonation upon electron transfer occurs on the side chain.[14] It would then be possible to obtain couplings through the side chain:

References

1 G. Sundholm, *Acta Chem. Scand.,* **25,** 3188 (1971).

2 M. Finkelstein and S. D. Ross, *Tetrahedron,* **1972**(28), 4497.

3 E. A. Mayeda, L. L. Miller, and J. F. Wolf, *J. Am. Chem. Soc.,* **94,** 6812 (1972).

4 M. Fleischmann, K. Korinek, and D. Pletcher, *Chem. Soc., Perkins Trans.,* **2,** 1396 (1972; *J. Electronal. Chem.,* **31,** 39 (1971).

5 D. A. White and J. P. Coleman, *J. Electrochem. Soc.,* **125,** 1401 (1978).

6 T. Shono, A. Ikeda, and Y. Kimura, *Tetrahedron Lett.,* **1971,** 3599.

7 A. Scott, P. A. Dodson, F. McCapra, and M. B. Meyers, *J. Am. Chem. Soc.,* **85,** 3702 (1963).

8 J. M. Bobbit, J. T. Stock, A. Marchand, and K. H. Weisgraber, *Chem. Ind.* (London), **1966,** 2127.

9 A. Rieker, E. L. Dreher, H. Geisel, and M. H. Khalifa, *Synthesis,* **1978**(11), 851.

10   R. A. Nash, D. M. Skauen, and W. C. Purdy, *J. Am. Pharm. Assoc.,* **47,** 433 (1958).

11   F. H. Covitz, French Patent 1,544,350 (1968).

12   T. Shono, Y. Yoshihiro, J. Hayashi, and M. Mizoguguchi, *Tetrahedron Lett.,* **1979**(2), 165.

13   J. Kaulen and H. J. Schafer, *Synthesis,* **1979**(7), 513.

14   R. C. Hallcher and M. M. Baizer, U.S. Patent 4,101,392 (1978).

2.3   Carboxylic Acids

Faraday first observed the evolution of gaseous products in the electrolysis of acetate solutions. Later, Kolbe[1] studied the electrochemical oxidation of carboxylic acids. After numerous studies through the years it has been established that the Kolbe reaction gives products by both radical and carbonium ion mechanisms.[2] The radical mechanism leads to alkanes, alkenes, and dimers, and is a one-electron oxidation. The carbonium ion, a two-electron step, leads to alkenes, alcohols, ethers, esters, ketones, and lactones. As with most types of electrolytic reactions the mechanistic pathways depend on several factors, most important of which are the anode material, the pH and nature of the medium, and the concentration of the carboxylic acid or acids.

The electrolysis is usually carried out in solutions of alkali metal salts of carboxylic acids. At platinum anodes the reaction proceeds preferentially as follows:

$$2RCO_2^- \xrightarrow{-2e} 2R^{\cdot} + 2CO_2$$
$$2R^{\cdot} \longrightarrow RR$$

This mechanism for the *Kolbe* coupling reaction is most widely accepted. When a solution of 5 N aqueous acetic acid is electrolyzed at a platinum anode, the Kolbe reaction takes place with almost 100% current efficiency.[3] However, when carbon is used as anode the course of the reaction changes. For example, instead of the reaction as shown above, methyl acetate was obtained in 82% along with some ethane, methane, and CO_2.[4] These results suggest that at carbon anodes the mechanism of the reaction is different from that at platinum anodes. The evidence is that at carbon anodes the deelectronation advances to the carbonium ion stage to give carbonium derived products:

$$R \xrightarrow{-e} R^{\dot{+}} \xrightarrow[-H^{\cdot}]{-e} R^+ \longrightarrow \text{products}$$

while at platinum free radicals are the preponderant primary electrodic products. In general, acidic media favor formation of radicals and hence coupling products, while alkaline media tend to favor carbonium ion and the products thereof. The Kolbe synthesis (the only electroorganic synthesis that rates a page in some organic textbooks) requires that about 2% of the carboxylic acid be present as a carboxylate ion.[5]

An interesting example of the effect of the electrolysis medium in the Kolbe reaction is the preparation of straight chain hydrocarbons from fatty acids, such as capric acid.[6] The yield of the dimer Kolbe product was related to the composition of the medium H_2O/CH_3OH. Depending on the water-alcohol ratios yields between 24 and 90% were obtained at platinum and graphite anodes. It is obviously very important to consider the composition of the medium in Kolbe-type electrolytic preparations, as is pointed out in critical reviews on the subject.[7] Of course, this is only one example of the effect of the electrolysis medium; its importance in synthetic work is general.

Dimer formation is favored in concentrated solutions. In dilute solutions of carboxylate at low current densities the cation formed at the anode finds it easier, statistically, to react with the solvent or some other electrophile than with another similar cation. It has been observed that, at low concentrations of carboxylate ion in a methanolic medium, the only product of the electrolysis is the methoxy derivative[8]:

$$R\left\langle\bigcirc\right\rangle CO_2^- \xrightarrow[-CO_2]{-2e} R\left\langle\bigcirc^+\right\rangle \xrightarrow[-H^+]{MeOH} R\left\langle\bigcirc\right\rangle OMe$$

The chemical reaction of the formed cation with the methanol solvent probably takes place in the bulk solution.

Cross-couplings of alkanoates are possible by anodic coelectrolysis[9]:

$$\left.\begin{array}{c} RCO_2^- \\ R'CO_2^- \end{array}\right| \xrightarrow[-2CO_2]{-2e} RR'$$

It has been possible in alkaline aqueous media to obtain alcohols by anodic oxidation of carboxylates substituted at the α-position by electrodotic groups[10]:

$$RCHXCO_2^- \xrightarrow[-CO_2]{OH^-} RCHXOH + 2e$$

Anodic oxidation of carboxylates can afford not only coupled products but also other products, depending on the electrolytic system as a whole. It has been shown, for example, that aryl acetates are oxidized according to these schemes[11]:

$$ArCH_2CO_2^- \xrightarrow[-CO_2]{-e} Ar\dot{C}H_2 \xrightarrow[-H^-]{O} ArCHO$$

$$Ar\dot{C}H_2 \xrightarrow{-e} Ar\overset{+}{C}H_2 \xrightarrow[-H^-]{MeOH} ArCH_2OMe$$

$$\downarrow$$

$$-\tfrac{1}{2}ArCH_2CH_2Ar$$

Coelectrolysis of a monocarboxylate and a half ester provides a way of extending the carbon chain of carboxylic acids, thus making possible the synthesis of fatty acids.[12]

Anodic oxidation of some long chain carboxylic acids, such as octanoic and 5-methylhexanoic acid, in fluorosulfuric acid has been found to afford lactones and unsaturated cyclic ketones.[13] In such a strong acidic medium electrons are removed from a carbon-hydrogen bond rather than from the usual bonds of the carboxyl group. The probable mechanism has been proposed as

$$\rangle\!\!-\!(CH_2)_n CO_2H_2^+ \xrightarrow[-H^-]{-2e} \rangle\!\!-\!\!\overset{+}{}(CH_2)_n CO_2H_2^+$$

$$\Big\Updownarrow OSO_2F^-$$

It seems as if the protonated (positive) part of the acid prefers to be away from the strongly positive anode surface, thus favoring removal of electrons from C—H bonds and expulsion of protons therefrom. In the usual Kolbe reaction the situation is, electrically speaking, different, the carboxyl group being negative and hence more prone to deelectronation than any other part of the molecule.

An interesting case of anodic decarboxylation reaction is the formation of uracil derivatives.[14] The electrolytic method is much more convenient than the thermal decarboxylation method and gives high yields of uracil:

It would be interesting, here again, to study this reaction in fluorosulfuric acid. In the basic medium a mechanism might be visualized as

In the fluorosulfuric acid a lactone may be expected to form:

References

1   H. Kolbe, *Ann. Chem. Pharm.*, **69**, 257 (1849).

2   L. Eberson, in S. Patai, Ed., *The Chemistry of Carboxylic Acids and Esters,* Wiley, New York, 1969, p. 211; G. E. Hawkes, J. H. P. Utley, and G. B. Yates, *J. Chem. Soc., Perkin II*, **1976**, 1709; J. H. P. Utley and G. B. Yates, *J. Chem. Soc., Perkins Trans.*, **1978** (4), 395.

3   K. Sugino, T. Sekina, and N. Sato, *Electrochem. Technol.*, **1**, 112 (1963).

4   K. J. Koel, Jr., *J. Am. Chem. Soc.*, **86**, 4686 (1964).

5   S. Wawzonek, *Synthesis*, **1971** (11), 285.

6   V. G. Gurger, *J. Appl. Electrochem.*, **8**, 207 (1978).

7   V. G. Gurger, *Chem. Ind. Development*, **10**, 15 (1976).

8   J. H. P. Utley and G. B. Yates, *J. Chem. Soc., Perkins Trans.*, **1978** (4), 395.

9   W. S. Greaves, R. P. Linstead, B. R. Shephard, S. L. S. Thomas, and B. L. C. Weedon, *J. Chem. Soc.*, **1950**, 3326.

10 H. Hofer and M. Moest, *Ann. Chem.*, **323**, 284 (1902).

11 J. P. Coleman, J. H. P. Utley, and B. C. Weedon, *J. Chem. Soc. (Chem. Commun.)*, **1971**, 438.

12 D. G. Bounds, R. P. Linstead, and B. C. Weedon, *J. Chem. Soc.*, **1954**, 448.

13 D. Pletcher and C. Z. Smith, *J. Chem. Soc., Perkin I*, **1975**, 948.

14 K. Matsumoto, H. Horikawa and T. Iwasaki, *Chem. Ind.* (London), **1978**, 920.

2.4 Substitutions

We have already discussed some anodic substitutions. Almost all types of substitutions are possible by the anodic method: acetoxylations, acetamidations, cyanations, hydroxylations, alkoxylations, halogenations, nitrations, thiocyanations, formoxylations.[1-6] The organic substance is electrolyzed in the presence of the appropriate anion, which is usually the anion of the supporting electrolyte, in a suitable solvent. The overall reaction can be expressed as

$$RE + N \longrightarrow RN + E^+ + e$$

where E and N denote electrophile and nucleophile. In most cases E is a proton while N may be any of the following nucleophiles: H_2O, OH^-, ROH, RO^-, CN^-, $RCOO^-$, SCN^-, CH_3CN, NO_3^-, pyridine, amines, NO_2^-, OCN^-, Cl^-, Br^-, I^-, F^-, or any other negative species that could be formed *in situ* electrolytically. If the species is transitory, reaction with it will have to take place at the surface of the electrode, while if of longer duration, the reaction could take place homogeneously and also at the surface of the electrode. If the species N is adsorbed at the electrode and reaction occurs at the adsorbed state, or heterogeneously, the possibility of obtaining stereochemical products and selectivities with polyfunctional molecules would be enhanced (direct substitutions on C—H bonds to form C—O or C—N bonds are not as easy by the usual chemical methods).

Anodic substitutions can be direct or indirect. Direct anodic substitutions proceed by deelectronation of the organic molecule whereby a radical cation forms and then reacts chemically with the nucleophile:

$$ArCH_3 \xrightarrow[-H^\cdot]{-2e} ArCH_2^+ \xrightarrow[-H^\cdot]{H_2O} ArCH_2OH$$

$$Cl^- \xrightarrow{-e} Cl^\cdot \xrightarrow{ArH} ArCl + H^\cdot$$

Anodic substitutions resulting in carbon-oxygen and carbon-nitrogen bond formation can be represented as follows:

$$ArCH_3 \xrightarrow[-H^+]{-e} Ar\overset{.}{C}H_2 \xrightarrow{-e} Ar\overset{+}{C}H_2$$

$$Ar\overset{+}{C}H_2 + CH_3OH \xrightarrow[-H^+]{} ArCH_2OCH_3$$

$$Ar\overset{+}{C}H_2 + CH_3CN \longrightarrow ArCH_2N\overset{+}{C}CH_3$$

$$ArCH_2N\overset{+}{C}CH_3 \xrightarrow[-H^+]{H_2O} ArCH_2NHCOCH_3$$

For the acetamidation reaction addition of water must be preceded by addition of acetonitrile.

A very interesting case illustrating the effect of the kind of electrolyte ions is the acetamidation of hexamethylbenzene[1]:

$$Ar(CH_3)_6 \xrightarrow[\substack{H_2O \\ Bu_4NClO_4}]{CH_3CN} Ar(CH_3)_5CH_2OH$$
$$5\%$$
$$+ Ar(CH_3)_5CH_2NHCOCH_3$$
$$95\%$$

When tetrafluoroborate salt is used instead of tetrabutylammonium perchlorate, the product distribution is reversed. It is seen that a mere change in the ions of the electrolyte changes drastically the distribution of the reaction products. This experimental fact, although very interesting, should not be very surprising. We must always remember the heterogeneous nature of electrochemical reactions and the existence and (unpredictable) influence of the structure of the electrical double layer. The ions of the supporting electrolyte are the building blocks of the electrified interfacial structure between electrode and solution. The course of the chemical reactions taking place at this interface is influenced by the composition of this part of the electrolysis medium. In this case, although the ratio CH_3CN/H_2O is the same in the bulk of the medium in both reactions, this ratio may not be the same at the double layer region of the anode when the perchlorate ion is replaced by the tetrafluoroborate ion. Not only the size but also the degree of hydration of the ions may be inherently different and further modified by the electric field. Thus with the BF_4^- ion a more hydrous atmosphere might be possible at the electrode surface than would be possible with the ClO_4^- ion. That possibility alone might be sufficient cause for the observed product distributions. Another possibility could be simply the ionic size differences, since such differences might influence the compactness of the electrical double layer and hence the potential gradient within this region. The hydrophilicity (water population) of the region may then be affected.

Another representative case of anodic substitution is a cyanation reaction:

$$RH \longrightarrow RH^{\ddagger} + e$$
$$RH^{\ddagger} \longrightarrow R^{\cdot} + H^{+}$$
$$R^{\cdot} \longrightarrow R^{+} + e$$
$$R^{+} + CN^{-} \longrightarrow RCN$$

An example of direct cyanation in acetonitrile/cyanide medium is the following[5]:

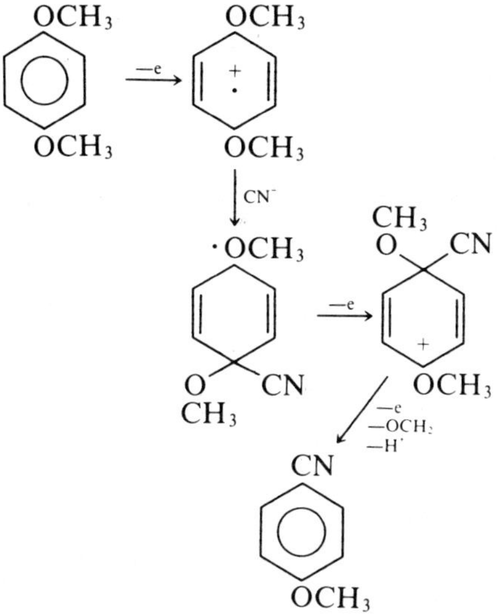

Direct anodic hydroxylation of aromatics is difficult. However, in trifluoroacetic acid as solvent with sodium trifluoroacetate or tetraalkylammonium salts as electrolytes, aromatic hydroxylations have been found to be possible. The produced phenols are more electroactive than the starting compounds, but in trifluoroacetic acid trifluoroacetate esters are formed that can then be readily hydrolyzed to give the phenol compound[6]:

$$ArH \xrightarrow[\text{TFA}]{-e} ArO_2CCF_3 \xrightarrow{H_2O} ArOH$$

This obviously requires that chemical reaction of the phenolic compound with the trifluoroacetic acid be very fast.

Substitutions by azide and cyanide groups have been achieved via anodic oxidation of N_3^- and CN^- at platinum anodes in acetonitrile containing azide or cyanide ions.[7,8]

Electrochemical thiocyanation of styrene in aqueous-acetic acid medium containing NH_4SCN and using platinum anode gives 1-phenyl-1, 2-bis(thiocyano)ethane.[9]

References

1　K. Nyberg, *Chem. Comm.,* **1969** (774).

2　K. Koyama, T. Susuki, and S. Tsutsumi, *Tetrahedron,* **1967** (23), 2675.

3　V. D. Parker and B. E. Burgert, *Tetrahedron,* **1965** (23), 4065.

4　L. Eberson and S. Nilsson, *Discuss. Faraday Soc.,* **45**, 242 (1968).

5　S. Andreades and E. W. Zahnow, *J. Am. Chem. Soc.,* **91**, 4181 (1969).

6　L. L. Miller, *Pure Appl. Chem.,* **51**, 2125 (1979).

7　G. Cauquis and G. Pierre, *Tetrahedron,* **1978** (34), 1475.

8　G. Cauquis and D. Serve, *Bull. Soc. Chim. Fr.,* **1979** (3–4, Pt. 2), 145.

9　O. V. Belyi, N. E. Karlovskaya, and L. M. Belaya, *Electrokhimiya,* **13**, 453 (1977).

2.5　Halogenations

Syntheses with electrogenerated halogens are a very appealing field of study and seem to have practical value. Substitutions by halide can be effected, as can other types of oxidations by anodically formed halogen.

With the exception of fluorinations (HF media) most anodic halogenations are indirect electrochemical reactions.[1] The halide ion is first deelectronated at the anode and the radical thus formed attacks the organic substrate:

$$X^- \xrightarrow{-e} X^{\cdot} \xrightarrow{R} RX^{\cdot} \longrightarrow \text{products}$$
$$2X^{\cdot} \longrightarrow X_2 \xrightarrow{RH} RX + HX$$

This type of mechanism is similar to the familiar photochemical mechanism of creating free halogen radicals.

The possibility of direct halogenations, as visualized below, does, however, exist:

$$R \xrightarrow{-e} R^{\ddagger} \xrightarrow{-e} R^+$$
$$R^+ + X^- \longrightarrow RX$$

Anthracene and naphthalene in acetonitrile containing tetraethylammonium bromide have been brominated by what seems to be a direct organic cation radical mechanism.[2]

The anodic preparation of propylene oxide involves an indirect anodic chlorination step. At the anode compartment the reactions are

$$2Cl^- \xrightarrow[H_2O]{-2e} HOCl + HCl$$

$$CH_3CH=CH_2 + HOCl \longrightarrow CH_3CHOHCH_2Cl$$

At the cathode compartment these reactions occur:

$$CH_3CHOHCHCl + OH^- \longrightarrow CH_3\overset{O}{\overset{\diagup \diagdown}{CH-CH_2}}$$

$$+H_2O + Cl^-$$

$$2H_2O + 2e \longrightarrow H_2 + 2OH^-$$

Actually, this is an example of homogeneous electrocatalysis, the catalyst being the chloride ion. The overall cell reaction, therefore, is the sum of the above reactions:

$$CH_3CH=CH_2 + H_2O \longrightarrow CH_3\overset{O}{\overset{\diagup \diagdown}{CH-CH_2}} + H_2$$

This reaction has been used on an industrial scale.[3]

Amines and amides can have their N—H bond transformed to N—X bond by halogen produced electrolytically.[4,5] This would be expected by analogy with chemical methods of forming N—X bonds.

Electrolysis of an aqueous solution of ammonium hydroxide and acetone containing potassium iodide gives a pyrazine and iodoform:

$$CH_3COCH_3 \xrightarrow{-e} \quad + HCI_3$$

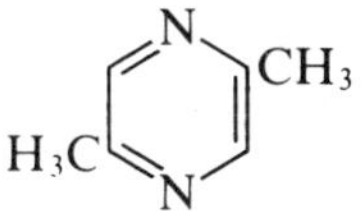

The reaction is apparently made possible by the *in situ* generation of NI$_3$.[6]

Halogenations of hydroxy compounds where the O—H bond can be converted to an O—X bond are possible. For example, *tert*-butyl alcohol was converted to the hypochlorite analog (CH$_3$)$_3$COCl in 79% yield.[4]

Usually, alcohols and phenols are oxidized to halogenated aldehydes or to polyhalides.[7] Ethanol, for instance, in basic solutions containing chloride is converted to chloroform and in acidic media to chloral. Ketones may give α-haloketones.

Indirect electrochemical oxidation of malonate esters to ethane-1,1,2,2-tetracarboxylate esters has been achieved in systems containing sodium iodide in acetonitrile or alcoholic solvents. The overall cell reaction is expressed as

$$2CH_2X_2 \xrightarrow[\text{I}^-]{-e} CHX_2CHX_2 + H_2$$

where X is—$COOCH_3$,—$COOC_2H_5$.

This is an electrocatalytic reaction, the iodide being the catalyst. Electrolysis of primary alcohols in the presence of anhydrous hydrogen halides affords acetals of 2-haloaldehydes.[8]

Unsaturated carbon-carbon bonds are subject to anodic halogenations, as would be expected. Ethylenic and acetylenic compounds in acetonitrile and dimethyl formamide afford chloroacetamides in acetonitrile, and chloroformates in dimethyl formamide.[9] In these electrolyses the solvents participate chemically in the overall electrochemical reaction, giving their characteristic products. If mixed nucleophilic organic solvent media are used for the electrolysis of polyunsaturated hydrocarbons, various addition products might form:

In nonnucleophilic media polymeric products would be formed:

Electrochemical chlorination of hydrolytic lignin has been feasible at graphite and platinum anodes.[10]

Finding suitable anodes for halogenations is not easy because nascent halogens attack most materials. The solvent itself may also be halogenated. Graphite and platinum anodes would be most suitable in most cases. Silver anodes form insoluble silver chloride and may be suitable in some aqueous chloride media. Nickel and platinum anodes are commonly used for fluorinations. A very interesting fluorination study has been reported recently by Drakesmith and Hughes[11] where octanoyl chloride was fluorinated to give perfluorooctanoyl fluoride and perfluorocyclic ethers.

References

1 J. Burdon and J. Tatlow, *Advances in Fluorine Chemistry,* Vol. 1, Academic, New York, 1960.

2 J. P. Millington, *J. Chem. Soc. (B),* **8,** 982 (1969).

3 J. A. M. Leduc, P. Konrad, and G. Troemel, Ger. Offen. 2,020,590 (1970).

4 E. J. Matzner, U.S. Patent 3,449,225 (1969).

5 M. Lamchen, *J. Chem. Soc.,* **1950,** 748.

6 S. H. Wilen and A. W. Levine, *Chem. Ind.,* **1969,** 237.

7 D. A. White, *J. Electrochem. Soc.,* **124,** 1177 (1977).

8 D. A. White and J. P. Coleman, *J. Electrochem. Soc.,* **125,** 1401 (1978).

9 M. Verniette, C. Daremon, and J. Simonet, *Electrochim. Acta,* **23,** 929 (1978).

10 E. I. Kovalenko, V. N. Shalimor, and V. A. Smirnov, *Zn. Prikl. Kim.* (USSR), **52,** 1312 (1979).

11 F. G. Drakesmith and F. G. Hughes, *J. Appl. Electrochem.,* **9,** 685 (1979).

2.6 Amines and Amides

Both aliphatic and aromatic amines are readily oxidized by the electrolytic method:

$$\begin{matrix} H \\ RN: \\ H \end{matrix} \xrightarrow{-e} \begin{matrix} H \\ RN^{\ddagger} \\ H \end{matrix} \longrightarrow \text{products}$$

Resonating structures can be written, such as

Aliphatic amines may suffer carbon-nitrogen bond cleavages upon anodic oxidation[1]:

$$(CH_3)_3N \xrightarrow[H_2O]{-2e} (CH_3)_2NH + CH_2O + 2H^+$$

$$R_2NCH_2R \longrightarrow R_2NH + RCHO$$

Anodic oxidation mechanisms of aromatic amines have been postulated as

Depending on the electrolysis medium, several reactions may be possible[2]:

In acetonitrile-pyridine media anilines have been oxidized to azobenzenes[3]:

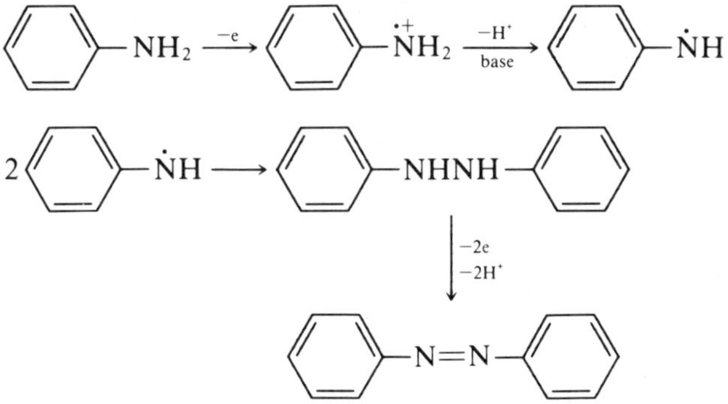

Oxidations of anilines and derivatives thereof can be carried out in water as dispersions or in mixtures of water with dimethylformamide, and with electrolytes such as potassium hydroxide, sodium acetate, or other salts. The azo compounds are obtained from such oxidations.[4]

Anodic substitutions of aromatic amines can be direct or indirect. In the latter case an electrophilic reagent is generated anodically and it then reacts chemically with the amine. For example, a halide ion can be oxidized to a halogen or halogen radicals and these species can react chemically with the amine.[5,6]

As would be expected amides require more positive potentials than the amines. While amines are generally oxidized around 1 V versus SCE, amides require about 2 V. Secondary and tertiary amides are oxidized at less positive potentials than the primary amides.[7] It is likely that amines are more prone to be adsorbed favorably on the anode than the amides, a possibility that would aid the oxidation of amines.

Amides of the type shown below have given dicarboxylic acids upon anodic oxidations[8]:

$$CH_3CONH(CH_2)_3COOH \xrightarrow{-e}$$
$$\xrightarrow{\text{hydrolysis}} HOOC(CH_2)_2COOH$$

Anodic oxidation of some anilides has been carried out in trifluoroacetic acid-KHF_2 media. Dimers and low yields of trifluoromethylated products were obtained.[9]

The trifluoroacetic acid is believed to stabilize the free radicals formed and thus to enhance dimer formation.

Tertiary aliphatic and heterocyclic amines in sodium cyanide-aqueous methanol media undergo anodic cyanation to give α-cyano amines and enamines.[10] The amine is apparently oxidized before cyanide, and the cyanide reacts nucleophilically with the positive amine species.

The anodic pathways of a number of *o*- and *m*-substituted anilines in aqueous-acidic and -basic media has been shown to lead to various coupling products.[11] The products depend greatly on the composition of the electrolysis medium, especially the pH. Although quantitative predictions of such solvent effects cannot be made, some qualitative predictions can be made, especially when the organic substrate is known to be inherently a basic or acidic molecule:

References

1   M. Masui and H. Sayo, *J. Chem. Soc.*, **1968**, 973; *J. Chem. Soc.*, **1971**, 1593.

2   S. Andreades and E. Zahnow, *J. Am. Chem. Soc.*, **91**, 4181 (1969).

3   S. Wawzonek and T. McIntyre, *J. Electrochem. Soc.*, **114**, 1025 (1967).

4   S. Wawzonek and T. McIntyre, *J. Electrochem. Soc.*, **119**, 1350 (1972).

5   G. Cauquis and G. Pierre, *C. R. Acad. Sci.*, **272**, 609 (1971).

6   K. Yoshida and T. Fueno, *J. Org. Chem.*, **37**, 4145 (1972).

7   J. F. O'Donnel and C. K. Mann, *J. Electroanal. Chem.*, **13**, 157 (1967); *J. Electroanal. Chem.*, **13**, 163 (1967).

8   S. Mizuno, *J. Electrochem. Soc. Jap.*, **29**, E27 (1961).

9   U. Hess, T. Gross, and B. Jahn, *Z. Chem.*, **19**, 25 (1979).

10   T. Chiba and Y. Takata, *J. Org. Chem.*, **42**, 2973 (1977).

11   R. L. Hand and R. F. Nelson, *J. Electrochem. Soc.*, **125**, 1059 (1978).

2.7 Ethers

Aliphatic ethers, in general, are very inert electrochemically. This makes them good solvents for electrolytic reactions, but they must be mixed with solvents of high dielectric constant to increase their ionizing ability for the supporting electrolyte. Aromatic ethers are oxidized at the anode.[1,2] When ethers are oxidized, the primary anodic product is a radical cation that is then further oxidized to a cation:

$$R-CH_2-O-CH_3 \xrightarrow{-e} R-CH_2-O-\overset{\cdot+}{C}H_3 \longrightarrow$$

$$\xrightarrow[-H^{\cdot}]{-e} R-CH_2-O-\overset{+}{C}H_2 \longleftrightarrow R-CH_2-\overset{+}{O}=CH_2$$

The oxidation of α,α-dimethoxystilbene in methanolic potassium hydroxide gave the tetramethoxy analog and a smaller amount of orthobenzoate.[3] Aromatic methoxy ethers can undergo various anodic substitutions, as for example substitution of aromatic hydrogen or methoxy groups by cyano groups[4]:

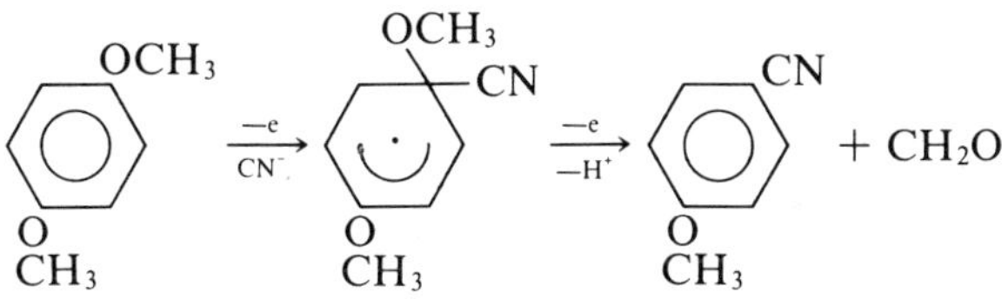

Intramolecular anodic cyclizations of methoxybibenzyls in acetonitrile or in media containing trifluoroacetic acid (TFA) (this acid is known to stabilize cation radicals) have been reported[5]:

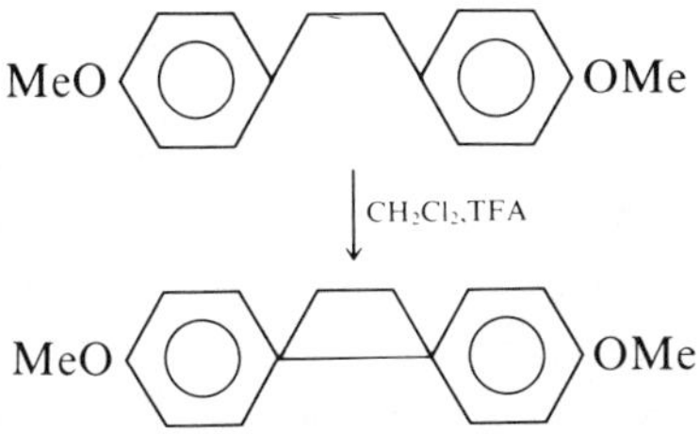

The corresponding cation radical was also formed, which, upon cathodic reduction, gave the dihydrophenanthrene.

According to Henton et al[6] the anodic addition of methanol to 1,4-dimethoxy aromatics to give the quinone bisketals takes place quite easily. The anodic oxidation method is, in their view, the method of choice for the preparation of compounds such as

References

1   A. Streitwieser, Jr., *Prog. Phys. Org. Chem.*, **1**, 1 (1963).

2   A. Zweig, A. H. Maurer, and B. G. Roberts, *J. Org. Chem.*, **32**, 1322 (1967).

3   R. Couture and B. Belleau, *Can. J. Chem.*, **50**, 3424 (1972).

4   S. Andreades and E. W. Zahnow, *J. Am. Chem. Soc.*, **91**, 4181 (1969).

5   A. Ronlan, O. Hammerich, and V. D. Parker, *J. Am. Chem. Soc.*, **95**, 7132 (1973).

6   D. R. Henton, R. L. McCreery, and J. S. Swenton, *J. Org. Chem.*, **45**, 369 (1980).

2.8   *N*-Heterocyclics

Because of the natural abundance and biochemical importance of *N*-heterocyclics, a great amount of work is being done with these substances.[1] The simple *N*-heterocycle, pyridine, under controlled conditions gives the dimeric salt in an acetonitrile-perchlorate medium[2]:

In a methanolic potassium hydroxide medium 2,6-dimethoxypyridine gave a mixture of tri-, tetra-, and pentamethoxy analogs.[3] *N*-methylpyrrole gave the 2,2,5,5,-tetramethoxy pyrroline.

Natural *N*-heterocycles, such as alkaloids, have received considerable attention.[4] At a carbon anode in acetonitrile, carbon-carbon and carbon-oxygen couplings have been reported.[5] The interesting observation was made that the C—C or the C—O—C coupling reactions are influenced greatly by the availability of the nitrogen lone electron pair (acid solution of N-acetyl analogs). The C—C coupling occurs when the lone pair is chemically bound, while C—O—C dimers are favored when the nitrogen pair is free (neutral media).

In general, the anodic oxidations (and also cathodic reductions) of this class of substances are very complex, especially when various functional groups exist in the N-heterocyclic nucleus. An idea of the complexity, and also of the potentials for preparative purposes, might be obtained from these examples[6]:

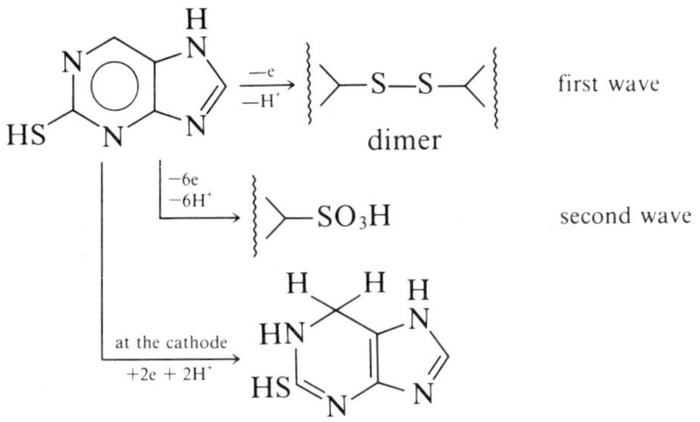

The preparative electrolyses shown were carried out at constant electrode potentials selected from polarographic current–potential curves.

Intramolecular nitrogen-sulfur bonds can be formed.[7] The anodic cyclization of N-(2-pyridyl)-thiobenzamide at controlled potential has given 2-phenyl-1,2,4-thiadiazolo [2,3-a] pyridinium perchlorate:

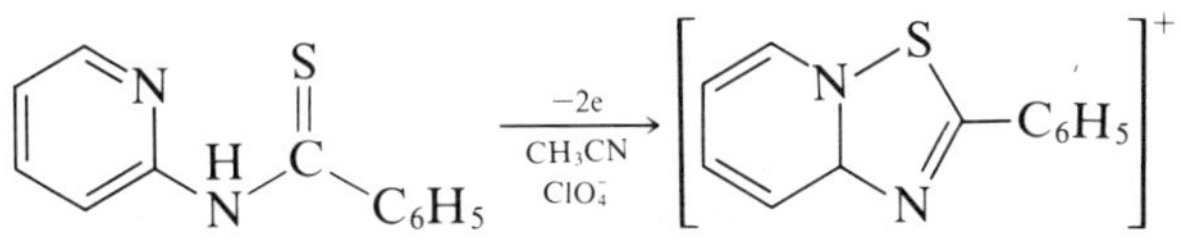

Anodic cyanation of N-alkylpiperidine and pyrrolidine derivatives has given two isomers with substitutions on the ring and on the side chain.[8] The final products seemed to depend more on steric effects than on the electrode potential.

Anodic oxidation of phenothiazines upon hydrolysis affords dihydroxy derivatives.[9] Neptune and McCreery proposed the following scheme:

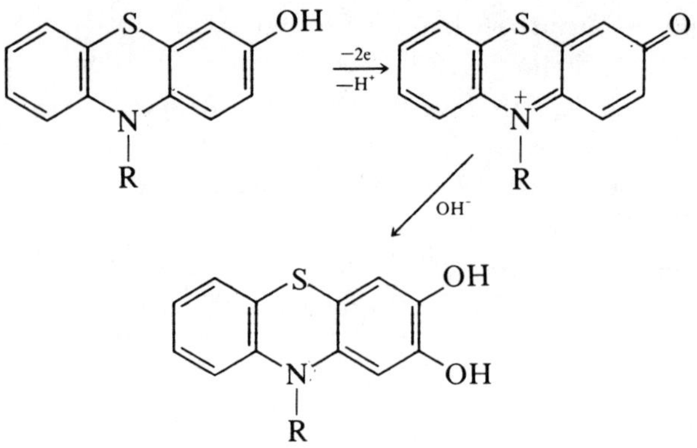

Heterocyclic amidines (precursors of biologically active compounds) are electrochemically quite active substances. As would normally be expected, the —NH$_2$ electrophore is deelectronated at less anodic potentials than the ring nitrogen function in these compounds. Their polarographic behavior suggests the possibility of modifying these molecules electrochemically on a preparative scale. Cauquis et al proposed the following one-electron mechanism for the anodic reaction of 2-amino-5-ethoxycarbonyl-4-methylthiazole in acetonitrile-lithium perchlorate medium[10]:

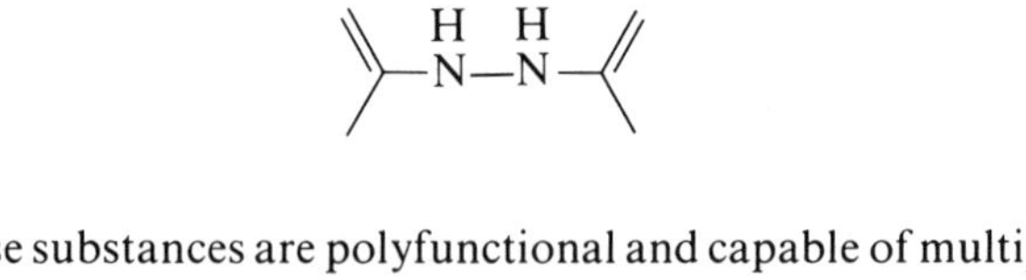

Because these substances are polyfunctional and capable of multiple-step electron transfers, it is necessary to perform the electrosynthesis under constant potential in order to attain maximum selectivity.

References

1   J. Volke, in A. R. Katritzky, Ed., *Physical Methods in Heterocyclic Chemistry,* Vol. 1, Academic, New York, 1963; S. Kwee and H. Lund, *Acta Chem. Scand.,* **23**, 2711 (1969); R. F. Nelson, in N. L. Weinberg, Ed., *Techniques of Electroorganic Synthesis,* Part II, Wiley-Interscience, New York; 1975, p. 269.

2 W. R. Turner and P. J. Elving, *Anal. Chem.*, **37**, 467 (1965).

3 N. L. Weinberg and H. R. Weinberg, *Chem. Rev.*, **68**, 449 (1968).

4 J. M. Bobbitt, J. T. Stock, A. Marchand, and K. H. Weisgraber, *Chem. Ind.* (London), **1966**, 2127.

5 J. M. Bobbitt, I. Noguchi, H. Yagi, and K. H. Weisgraber, *J. Am. Chem. Soc.*, **93**, 3551 (1971).

6 G. Dryhurst, *J. Electroanal. Chem.*, **28**, 33 (1970).

7 I. Tabakovic, M. Trkovnik, M. Batusic, and K. Tabakovik, *Synthesis*, **1979** (8), 561.

8 T. Chiba and Y. Takata, *J. Org. Chem.*, **42**, 2973 (1977).

9 M. Neptune and R. L. McCreery, *J. Medic. Chem.*, **22**, 196 (1979).

10 G. Cauquis, H. M. Fahmy, G. Pierre, and M. H. Elnagdi, *J. Heterocycl. Chem.*, **16**, 413 (1979).

CATHODIC REACTIONS

2.9 Formation of Carbon-Hydrogen Bonds

Organic compounds can be electrolytically hydrogenated directly or indirectly. Direct hydrogenations are preferably called *electroreductions*, implying that the organic substrate accepts electrons directly from the electrode:

$$R + e \longrightarrow R^{\cdot -} \quad \text{electrical step}$$

$$R^{\cdot -} + H^+ \longrightarrow RH^{\cdot} \quad \text{chemical step}$$

$$RH^{\cdot} + e \longrightarrow RH^- \quad \text{electrical step}$$

$$RH^- + H^+ \longrightarrow RH_2 \quad \text{chemical step (ECEC mechanism)}$$

Chemical reactions with a proton source follow or concur with the electron exchange event. It is obvious that any electrophile can attack the primary product $R^{\cdot -}$. For practical hydrogenation purposes, therefore, good protonic media, such as H_2O, and various aqueous-organic or organic-acid mixtures would be used. The ease of many electroreductions is a function of the prevailing pH of the solution (the pH at the surface of the electrode, strictly speaking). In some cases buffered solutions would preferably be employed. By proper pH and electrode potential control, selective reductions can be achieved that may be difficult or even impossible by conventional chemical methods.

Electroreductions of unsaturated systems can be effected by addition of electrons to unsaturated bonds:

$$R'CH{=}CHR'' + e \longrightarrow R'\dot{C}H{-}\overline{C}HR''$$
$$\downarrow {\scriptstyle H^\cdot}$$
$$R'\dot{C}H{-}CH_2R''$$
$$\downarrow {\scriptstyle e}$$
$$R'CH_2CH_2R'' \xleftarrow{\ H^-\ } R'\overline{C}H_2CH_2R''$$

The relative ease of the reduction depends on the influence of the R' or R'' on the unsaturated bond. Activated olefins can dimerize to give difunctional molecules. An example is the dimerization of acrylonitrile to give adiponitrile[1]:

$$2CH_2{=}CHCN + 2e + 2H^+ \rightarrow NC(CH_2)_4CN$$

Electroreductions of polyunsaturated hydrocarbons may afford intramolecular cyclizations[2]:

$$(CH_2)_n{\Big\langle}\begin{array}{l}CH{=}CHR'\\CH{=}CHR''\end{array} \xrightarrow[2H^\cdot]{2e} (CH_2)_n{\Big\langle}\begin{array}{l}CH{-}CH_2R'\\CH{-}CH_2R''\end{array}$$

where R' and R'' are electrophilic activating groups. Formation of polymeric products, especially in aprotic media, is possible with such materials.

When a hydrogenolysis is effected by the electrolytic method, it is called *electrohydrogenolysis*. Generally, electrohydrogenolyses are expressed as

$$RX + 2e \rightarrow R^- + X^-$$
$$R^- + H^- \rightarrow RH$$

Indirect hydrogenations are preferably called *electrohydrogenations*. They occur by chemical reaction of electrogenerated hydrogen with the organic substrate:

$$2H^+ + 2e \longrightarrow H_2 \text{ or } 2H^\cdot$$
$$H_2 + R \longrightarrow RH_2$$
$$2H^\cdot + R \longrightarrow RH_2$$

In general, the rate of hydrogenation by the electrolytic method increases with the acidity of the electrolysis medium.[3] In indirect hydrogenations the electrode surface plays the very important role of an electrocatalyst. This type of hydrogenation is similar to nonelectrochemical catalytic hydrogenations. However, the electric field and the particular structure of the double layer may influence the course of the reaction and hence the nature of the ultimate products.

Indirect hydrogenations are also possible by the *in situ* electrogeneration of a reducing agent, as, for example, when using the reversible redox couple[4]:

$$Sn^{2+} \rightleftharpoons Sn^{4+}$$

$$RCH_2X + Sn^{2+} \xrightarrow[H_2O]{} RCH_3 + Sn^{4+}$$

At the cathode we find:

$$Sn^{4+} + 2e \rightarrow Sn^{2+}$$

The stannous ion chemically reduces the organic molecule, thereby becoming oxidized to a stannic ion, which is reduced again at the cathode to repeat the cycle. Such indirect electrolytic reductions are homogeneous reactions. In principle only catalytic amounts of the reductant would be sufficient. By chemical methods at least an equivalent amount of reductant would be needed, and the problem of isolating the products and reactants may be difficult. It is apparent that this is a very promising area of research. Organic reversible redox systems might also be found for homogeneous electrocatalytic reactions. Andrieux et al tried the use of anthracene, phenanthridine, benzonitrile, and some other substances to produce reversible couples for catalytic reduction of halobenzenes and halopyridines.[5] A possible catalytic mechanism can be expressed as follows:

$$Ar + e \rightleftharpoons Ar^{\bar{\cdot}} \qquad \text{catalytic couple}$$

$$Ar^{\bar{\cdot}} + RX \longrightarrow R^{\cdot} + X^- + Ar$$

$$R^{\cdot} + e \longrightarrow R^-$$

$$R^- + H_2O \longrightarrow RH + OH^-$$

Indirect electrohydrogenations are generally performed with electrodes of low hydrogen overvoltage, namely, platinum, copper, nickel, iron, and alloys thereof, whereas electroreductions (as defined above) are generally carried out with electrodes of high hydrogen overvoltage, such as mercury, lead, zinc, and tin. Electrohydrogenations, being indirect reductions, cannot be as selective as electroreductions. Selective electroreduction of multiple double bonds can often be accomplished by constant potential electrolysis. A unique example is one reported by Ankner et al[6] using the hydrocarbon [2₄]paracyclophanetetraene.

Ph =phenyl

The double bonds were selectively hydrogenated by carrying out constant potential electrolysis at potentials chosen polarographically. Although all four olefinic bonds are initially equivalent, they cease to be so as soon as the first hydrogenation occurs. The remaining bonds are electroreduced by successively electrolyzing at more negative potentials.

In general, conjugated double bonds constitute more electroactive systems than single double bonds. Condensed aromatics are more easily electroreduced than the less condensed aromatics. Anthracene, for instance, is reduced before naphthalene and napthalene before benzene. In fact, benzene is very difficult to reduce directly (it can be reduced by solvated electrons).

The possibilities of transfer hydrogenations by electrochemically generating hydrogen donor substances, such as 9,10-dihydroanthracene, 1,4-dihydronaphthalene, and other such donors, deserve some consideration. Conversely, hydrogen acceptors may be produced electrolytically. For example, benzoquinone might be generated from hydroquinone, so that only catalytic amounts of the benzoquinone would be needed for a hypothetical reaction such as:

References

1 M. M. Baizer, U.S. Patent 3,193,408–1 (1963).

2 J. Petrovich, J. D. Anderson, and M. M. Baizer, *J. Org. Chem.*, **31**, 3897 (1966).

3 A. N. Frumkin, in P. Delahay and C. W. Tobias, Eds., *Advances in Electrochemistry and Electrochemical Engineering,* Vol. 3, Interscience, New York, 1963, p. 381.

4 H. C. Rance and J. M. Coulson, *Electrochim. Acta,* **14**, 283 (1969).

5 C. P. Andrieux, C. Blocman, J.-M.D. Bouchiat, and J.-M. Saveant, *J. Am. Chem. Soc.,* **101**, 3431 (1979).

6 K. Ankner, B. Lamm, B. Thulin, and O. Wenström, *Acta Chem. Scand.,* **33**, 391 (1979).

2.10 Carbonyl Compounds

The carbonyl group is an electrophoric group offering interesting synthetic possibilities. Electrolytic reduction of this function can be effected in both protic and aprotic media. In protonic media the reduction of ketones is generally expressed by the following scheme[1-4, 8]:

$$R_1COR_2 \xrightarrow{e} R_1\dot{\bar{C}}OR_2 \xrightarrow{H^{.}} R_1\underset{.}{\overset{OH}{C}}R_2 \longrightarrow \text{dimer}$$

$$\downarrow^{e}_{H^{.}}$$

$$R_1CHOHR_2$$

The ease of electronation is influenced by the nature of R_1 and R_2. Polarographic reductions of ketones may show one or two waves depending on the pH of the medium. In acidic solution two one-electron waves are usually formed. In basic media the waves merge into one two-electron wave that appears at more negative potentials than in acidic media. This implies that the proton is involved in the rate determining step of the electrodic reaction. In alkaline media alcohol formation is the predominant reaction, while in acidic media pinacols are the major products. If stereoisomers are possible, it is more likely that such products will be obtained by electrochemical reduction than by chemical methods.[5] As is the case with chemical reductions, aldehydic functions are reduced more easily than ketonic functions. Ester and carboxylic carbonyl groups are most difficult to reduce electrolytically, unless they are activated by electron withdrawing groups.

Reduction of carbonyls to a hydrocarbon state is generally very difficult. When it is possible it is most likely an indirect catalytic electrohydrogenation reaction, and would usually take place at low hydrogen overvoltage metals, which are known to catalyze hydrogenations by conventional methods.

In nonaqueous media [eg, tetrahydrofuran (THF)] some interesting cyclizations can be achieved[6]:

Quinones in aprotic media can be reduced stepwise:

Anthraquinone in the presence of ethylbromide gave the diethyl ether in 20% yield.[7] Elving and Leone studied the effect of pH and potential on the reduction of several phenyl ketones in aqueous-methanol media.[8] In their view the electrode polarizes the carbonyl group and thus prepares an easy path for the direct electron transfer from the electrode to the adsorbed organic molecule:

In this process the proton also contributes to the ease of the electron transfer, as implied above. This is an example of how the proton is involved in the primary rate determining step of the electrode reaction. (Although polarizations of organic molecules in the prevailing electric fields are always to be expected to varying degrees, the direction of polarization cannot be easily inferred because it may depend upon on which side of the *zero charge potential* the organic species is adsorbed or oriented at the electrode surface. Furthermore, if a molecule is adsorbed at an electrode when the electrode is at the zero charge potential, there will be no field-induced polarization.) The reduction of aldehydes would require divided cells, owing to their easy oxidation at the anode, unless an anodic depolarizer, which would be preferentially oxidized at the anode, is used.

In aprotic solvents the electrochemical reduction of ketones seems to be influenced by metal salts, such as $CoCl_2$, $ZnCl_2$, and $Ni(AcO)_2$. The reduction involves one electron and is reversible in the absence of the salts but irreversible in their presence.[9] Metallic ions in aprotic media may sometimes exert very peculiar effects, since they tend to form complexes with the organic molecules and also to modify the electrical double layer.

Anhydrides can be reduced in both aqueous and organic media. In aqueous media hydrolysis may be rapid. In aqueous-organic media the hydrolysis may be retarded so that electrolytic reduction can be possible.[10]

References

1   F. D. Popp and H. P. Schultz, *Chem. Rev.,* **62**, 29 (1962).

2   S. Wawzonek and H. Laitinen, *J. Am. Chem. Soc.,* **63**, 2341 (1941).

3   P. Zuman, *Collect. Czech. Chem. Commun.,* **33**, 2548 (1968).

4   P. Zuman, D. Barns, and A. Ryvolova-Kejharova, *Discuss. Faraday Soc.,* **45**, 202 (1968).

5   J. H. Stocker and R. M. Jenevein, *J. Org. Chem.,* **33**, 2145 (1968).

6   T. J. Curphey, C. W. Amelloti, T. P. Layoff, R. I. McCartney, and J. H. Williams, *J. Am. Chem. Soc.,* **91**, 2817 (1969).

7   S. Wawzonek, R. Berkey, E. W. Blaha, and M. E. Runner, *J. Electrochem. Soc.,* **103**, 456 (1956).

8   P. J. Elving and J. T. Leone, *J. Am. Chem. Soc.,* **80**, 1021 (1958).

9   A. Diaz, M. Parra, R. Banuelos, and R. Contreras, *J. Org. Chem.,* **23**, 4461 (1978).

10   D. Kyriacou, *Anal. Chem.,* **42**, 805 (1970).

2.11   Carboxylic Acids

Carboxylic acids can be reduced cathodically if they are activated by electron withdrawing groups in the molecule. They can then give aldehydes, alcohols, and, in some cases, hydrocarbons.[1] Acids that are not activated contribute only to an increase in current by increasing the acidity of the medium (hydrogen evolution). Aromatic carboxylic acids are easier to reduce than aliphatic acids, in general. Benzoic acid can be reduced to benzyl alcohol. Under special conditions the reduction of some aromatic carboxylic acids can be stopped at the aldehyde stage (divided cell)[2]:

$$ArCO_2H + 2e \xrightarrow[\substack{2H^{\cdot} \\ \text{(benzene)}}]{} ArCH(OH)_2 \xrightarrow{-H_2O} ArCHO$$

Apparently, the benzene extracts the aldehyde and thus retards its further reduction. There will be several cases where heterogeneous electrolysis media, like the one above, supress further reduction or oxidation of the organic substrate.

References

1 F. D. Popp and H. P. Schultz, *Chem. Rev.*, **62**, 19 (1962); P. E. Iversen and H. Lund, *Acta Chem. Scand.*, **21**, 389 (1967).

2 J. Wagenknecht, *J. Org. Chem.*, **37**, 1513 (1972).

2.12   Nitro Compounds

The nitro group is one of the best electrophores as regards both ease of reduction and versatility of derived products. As a class organic nitro compounds have been studied most extensively, especially the aromatics. A great variety of products can be obtained by electroreduction of these substances, depending on electrolysis conditions and the nature of the electrode material. Nitroalkanes are reduced in both aqueous and anhydrous media. In basic media the acianion, a negative species, is apparently very resistant to reduction. In acid solutions and at mercury or lead cathodes amines are mostly produced, while at low overvoltage cathodes the major products are hydroxylamines.

In anhydrous media the reduction of nitroalkanes apparently goes through a radical anion that is unstable and dissociates to give a free radical and a nitrite ion[1]:

$$RNO_2 + e \longrightarrow \overset{\displaystyle \overline{O}}{\underset{\displaystyle O}{\cdot RN}} \longrightarrow R^{\cdot} + NO_2^{-}$$

$$2R^{\cdot} \longrightarrow RR$$

The formation of these dimers is reminiscent of the anodic (Kolbe) formation of dimers from carboxylic acids. Aromatic nitrocompounds are also believed to be reduced by first forming radical anions[2]:

The advantage of the electrolytic method might be that it enables us to select the potential of the cathode (reducing power) so as to direct the course of the reaction toward the desired objective. The complexity and, at the same time, the versatility of electrochemical reduction of nitro compounds can be seen in the scheme below. All these products, of course, can also be obtained by chemical reductions. The mechanism of this reduction has been studied extensively.[2-5] Because the reduction products are susceptible to anodic oxidation, divided cells are required. Reductions are possible at all pH's but the products will depend on pH and electrode potential:

Reduction of aliphatic nitro compounds in acidic media with nickel, copper, and mercury electrodes affords hydroxylamines, which can be isolated as derivatives. With controlled potential electrolysis the alkylhydroxylamines of nitromethane, nitroethane, and 1- and 2-nitropropane have been obtained.[6]

Reductive acetylation of aliphatic and aromatic nitro and nitroso compounds has been carried out in aprotic media containing acetic anhydride.[7]

$$RNO_2 \xrightarrow[Ac_2O]{4e} RN\begin{matrix} OAc \\ \\ C=O \\ CH_3 \end{matrix} \xrightarrow[Ac_2O]{2e} RNO$$

References

1 H. Sayo, Y. Tsukitani, and M. Masui, *Tetrahedron,* **1968** (24), 1717.

2 B. Kastening, *Naturwissenschaften,* **47**, 443 (1960).

3 J. Pearson, *Trans. Faraday Soc.,* **44**, 683 (1948).

4 S. Wawzonek and J. O. Fredicson, *J. Am. Chem. Soc.,* **75**, 3985 (1955).

5 H. L. Piette, P. Ludwig, and R. N. Adams, *Anal. Chem.,* **34**, 916 (1962).

6 P. E. Iversen and H. Lund, *Acta Chem. Scand.,* **19**, 2303 (1965).

7 L. Christensen and P. E. Iversen, *Acta Chem. Scand.,* **B33**, 352 (1979).

2.13 Unsaturated Hydrocarbons

In general, the electrochemical reduction of isolated double bonds has been very difficult. When activating groups, such as $-COO$ and $-CN$, are situated close to the unsaturated bond, the reduction is facilitated greatly. Some reductions are believed to be accomplished by reaction of electrogenerated hydrogen atoms or molecules at the electrode surface, as was stated in Section 2.9, with the unsaturated compound (indirect reduction):

$$H^+ + e \longrightarrow \dot{H} \xrightarrow{-C=C-} -\dot{C}-\overset{H}{\underset{}{C}}-$$

Propylene, for instance, can be reduced to propane by such a mechanism.[1] Another reduction path is the direct transfer of electrons (electroreduction) from the cathode to the organic substrate, followed by addition of a proton or any electrophile capable of forming a bond with carbon:

$$\text{C=C} + e + H^+ \longrightarrow \overset{\cdot}{\text{C}}-\text{C}\overset{H}{\diagup} \overset{e, H^+}{\longrightarrow} \overset{H}{\text{C}}-\overset{H}{\text{C}}$$

$$\downarrow$$

dimer

If electrophiles are unavailable, polymerization would be expected, as is also implied above[2,3]:

$$CH_2{=}CCH_3CH_3 \xrightarrow[\text{THF}]{} \text{polyisoprene}$$

$$\text{styrene} \xrightarrow[\substack{R_4N^+ClO_4 \\ 0°C}]{} \text{polystyrene}$$

The electrochemical dimerization of acrylonitrile to adiponitrile on an industrial scale is at present the largest electroorganic process in the world (1 billion lb/y). The reactions have been described as follows[4]:

$$2CH_2{=}CHCN + 2e + 2H^+ \qquad \text{cathode}$$

$$\downarrow$$

$$NCCH_2CH_2CH_2CH_2CN$$

$$H_2O \longrightarrow \tfrac{1}{2}O_2 + 2H^+ + 2e \qquad \text{anode}$$

The electrolyte is tetraethylammonium p-toluene sulfonate. This salt is used in large amounts to solubilize the acrylonitrile.[5] It is believed that the large cation favors the dimerization reaction. This type of reaction is known as *hydrodimerization*. Such reactions are possible with variously activated olefins. Recently, it has been possible to use an undivided cell for the hydrodimerization of acrylonitrile using potassium phosphate as electrolyte and only a trace of Bu_4N^+ ion.[6] Apparently, even traces of the quaternary ion can participate in the formation of the electrical double layer structure, which is in favor of the hydrodimerization reaction. Effects of very small concentrations of charged species on electrochemical phenomena are frequently observed in various ways in polarographic studies.[7] It should not be surprising at all that small amounts of apparently inert ions sometimes influence drastically the course of electrochemical reactions. We may perhaps dispel any doubts about this if we recognize that a small amount of some species in the bulk solution may be a very large proportion if concentrated in the region of the electrical double layer. Intradimerizations of activated olefins can give cyclic products[8]:

Benzene, although not reducible polarographically, can be reduced by solvated electrons in both divided and undivided cells. The products from the two cells are different[9]:

References

1   H. J. Barger, Jr., *J. Org. Chem.,* **34,** 1489 (1969).

2   B. Funt, S. Bhadani, and D. Richardson, *Polymer Reprints,* **7,** 153 (1966).

3   M. Ohta, *Bull. Chem. Soc. Jap.,* **17,** 485 (1942).

4   Monsanto Co., *Chem. Eng.,* Nov. 8, 238 (1965).

5   J. P. Petrovich, J. D. Anderson, and M. M. Baizer, *J. Org. Chem.,* **31,** 3897 (1966); M. M. Baizer, *J. Electrochem. Soc.,* **111,** 215 (1964).

6   W. V. Childs and H. C. Walters, *AIChE Symp. Ser.,* **75,** 19 (1979).

7   D. Kyriacou, *Anal. Chem.,* **32,** 1893 (1960).

8   J. D. Anderson, M. M. Baizer, and J. P. Petrovich, *J. Org. Chem.,* **31,** 3890 (1966).

9   R. A. Benkeser and E. M. Kaizer, *J. Am. Chem. Soc.,* **85,** 2858 (1963); S. H. Langer and S. Yurchak, *J. Electrochem. Soc.,* **116,** 1228 (1969).

2.14   Carbon-Nitrogen Bonds

Inactivated alkyl nitriles are difficult to reduce by the electrolytic method. This relative inertness makes acetonitrile a good solvent for many electrolytic syntheses. It can partake, however, in the overall reaction by reacting chemically as a nucleophile with the primary electrodic products. In contrast aromatic nitriles are easily reduced. The initial electrode products are frequently stable radical anions and are formed reversibly. In aqueous media the corresponding amines can be obtained in high yields. In aprotic media [eg, dimethylformamide (DMF)] dimeric products can be formed[1]:

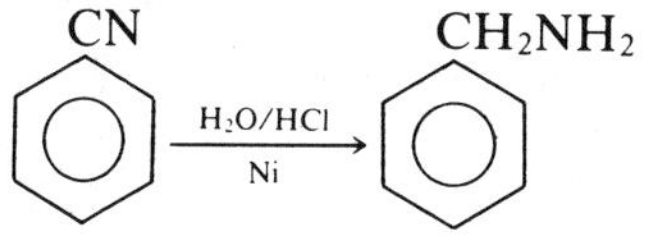

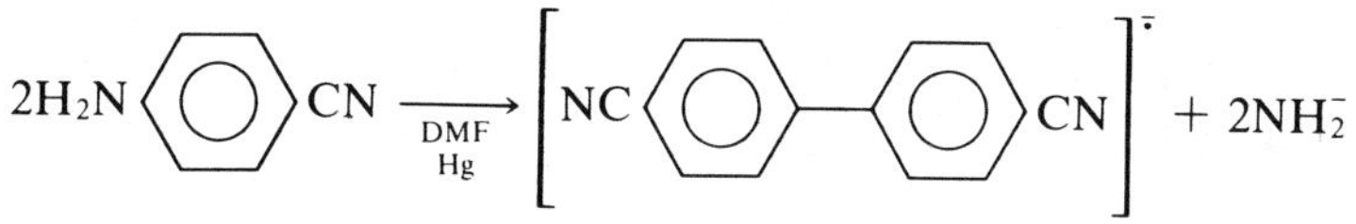

The mechanism of the reduction in aprotic media has been postulated to be a direct electronation of the π system of the nitrile group, giving a radical anion[2]:

$$\mathrm{ArCN} + e \qquad \mathrm{Ar\dot{C}{=}\overline{N}}$$

Imines and oximes are readily reducible in both acidic and basic media[3]:

$$>\mathrm{C{=}NH} + \mathrm{H^+} \rightleftharpoons\ >\mathrm{C{=}\overset{+}{N}H_2}$$

$$>\mathrm{C{=}\overset{+}{N}H_2} + e \longrightarrow\ >\mathrm{\dot{C}NH_2} \xrightarrow[\mathrm{H^+}]{e}\ >\mathrm{CHNH_2} \qquad \text{acidic media}$$

$$>\mathrm{C{=}NH} \xrightarrow[\mathrm{H_2O}]{e}\ >\mathrm{\dot{C}NH_2} + \mathrm{OH^-}$$

$$>\mathrm{\dot{C}NH_2} \xrightarrow[\mathrm{H_2O}]{e}\ >\mathrm{CHNH_2} + \mathrm{OH^-} \qquad \text{basic media}$$

Oximes give the corresponding amines. The reduction potentials, as we would expect, are dependent on pH. This suggests that protonation of the substrate precedes or concurs with electronation. (In general, if the potential, $E_{1/2}$, does not vary with pH, it would be reasonable to conclude that the proton is not involved in the primary electrodic reaction.)

Very interesting syntheses could be accomplished by electrochemical reduction of certain carbon-nitrogen bonds. Once a radical anion is formed various electrophiles can combine with the anion:

$$RCH{=}N{-} \xrightarrow{\ e\ } R\dot{C}H{-}\bar{N}{-}$$
$$\downarrow E$$
$$R\dot{C}H{-}NE{-}$$
$$\downarrow e, E$$
$$RCHENE{-}$$

Degrand et al demonstrated the possibility of preparing pyrrolidine and piperidine derivatives from Schiff bases and 1-ω-dibromoalkanes[4]:

$$ArCH{=}NAr \underset{\ }{\overset{e}{\rightleftharpoons}} Ar\dot{C}H{-}\bar{N}Ar$$
$$\downarrow Br(CH_2)_nBr$$
$$Ar\dot{C}HNAr$$
$$|$$
$$(CH_2)_nBr + Br^-$$
$$\downarrow e$$
$$Ar\bar{C}HNAr$$
$$|$$
$$(CH_2)_nBr$$
$$\downarrow {-}Br$$
$$ArCH{-}NAr$$
$$\diagdown(CH_2)_n$$

References

1   P. H. Rieger, I. Bernal, W. H. Reinmuth, and G. K. Fraenkel, *J. Am. Chem. Soc.*, **85,** 683 (1963).

2   B. L. Funt, *Macromolecular Reviews,* Vol. 1, Interscience, New York, 1967, p. 35.

3   H. Lund, *Acta Chem. Scand.,* **13,** 249 (1959).

4   C. Degrand, C. Grodemouge, and P. L. Compagnon, *Tetrahedron Lett.,* **1978,** 3023.

2.15   Sulfur Compounds

Organosulfur compounds have been extensively studied polarographically. With mercury cathodes certain sulfur compounds tend to react chemically to form organomercury compounds[1]:

$$RSSR + Hg \longrightarrow (RS)Hg$$

Homocystine, for example, is reduced to homocystein.[2] The resulting organometallic compound is electroreduced to the mercapto analog and free metal:

$$(RS)_2Hg + 2e + 2H^+ \longrightarrow 2RSH + Hg$$

Sulfones, upon reduction, give sulfinic acids and mercaptans.[3,4] Certain aromatic sulfones[5,6] and sulfonamides[7] have their carbon-sulfur bonds cleaved. Sulfoxides are reduced to sulfides:

$$\overset{O}{\underset{O}{RSR}} + 2e + 2H^+ \longrightarrow RSO_2H + RSH$$

Electroreduction of disulfides in aprotic media affords sulfinic acids.[8] The formation of thiolates in the presence of oxygen is the basis for sulfinic acid formation:

$$RS^\cdot + O_2^{\bar{\ }} \longrightarrow RSO_2^-$$

In general, the electrolytic reduction of single carbon-sulfur bonds is very difficult. Double carbon-sulfur bonds, however, are easily reduced.

References

1   J. Donahue and J. W. Oliver, *Anal. Chem.*, **41**, 753 (1969).

2   M. J. Allen and H. Steinman, *J. Am. Chem. Soc.*, **74**, 3932 (1952).

3   D. Barnard, M. B. Evans, G. Higgins, and F. Smith, *Chem. Ind.*, **1961**, 20.

4   R. Bowers and H. Russell, *Anal. Chem.*, **32**, 405 (1960).

5   D. Kyriacou, unpublished work.

6   J. Grimshaw and J. Trocha-Grimshaw, *J. Chem. Soc., Perkin I,* **1979,** 799.

7   O. Manovsck, D. Exner, and P. Zuman, *Collect. Czech. Chem. Commun.,* **33,** 4000 (1968).

8   C. Degrand and H. Lund, *Acta Chem. Scand.,* **B33,** 512 (1979).

2.16   Carbon-Halogen Bonds

Alkyl and aryl halides can be reduced electrochemically to their corresponding hydrocarbons. Two electrons are involved in the process.[1] In dimethylformamide one-electron reductions have been observed.[2] This would normally be expected in aprotic solvents.

It has been generally stated that the electroreduction of organic halides is independent of electrode material and pH. Such a broad generalization may be confusing rather than instructive. If selective reductions among several carbon-halogen bonds are desirable, the electrodes surface morphology, as well as its material, may be of the utmost importance. Furthermore, there are cases where the pH of the medium influences the reduction potentials. Various chloropyridine derivatives have been found to depend on pH and electrode material for the electrolytic reductions of carbon-halogen bonds.[3] Depending on conditions a great variety of products can be obtained by the electrochemical reduction of organic halides. Very reactive intermediates, such as benzynes and carbenes, have been obtained in aprotic media.[4] Polymeric products have also been obtained.[5]

The general reduction mechanism has been postulated as follows[6]:

$$\left[RX \xrightarrow{e} [R^{\delta^{\cdot}} \cdots X^{\delta}]^{\overline{\cdot}} \longrightarrow R^{\cdot} + X^{-} \atop \underset{RR \qquad\qquad RH}{\overset{R^{\cdot} \quad H^{\cdot}}{\swarrow \qquad \searrow^{e}}} \right]$$

In protonic media the hydrogenated product would be predominant, while in aprotic media free radical derived products would be expected as well as various substitutions and couplings, depending on the medium of electrolysis. Organometallic products may also be obtained by reaction with the cathode itself. The examples cited below have been chosen for the purpose of demonstrating the potential synthetic versatility of organic halides, using the electrochemical method[7]:

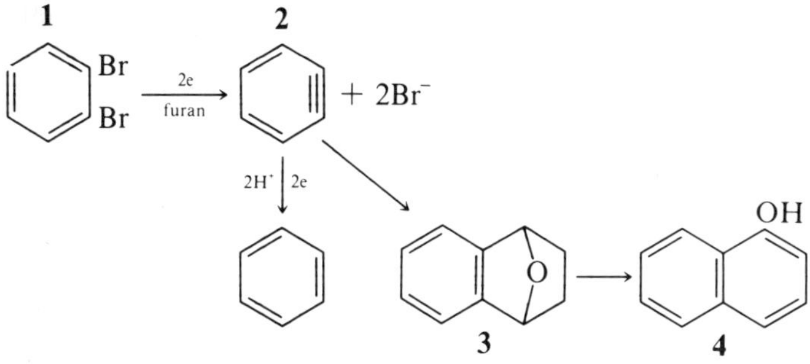

Although only a low yield of α-naphthol was isolated under the electrolysis conditions used, the possibility of discovering more favorable conditions for its formation exist. The relative yields of **3** and **4** depend on the stability of **2** in the medium. This would be influenced by both solvent and kind of electrolyte. Also, the rate of diffusion of the benzyne intermediate away from the electrode surface will determine the course of the reaction and the nature of the final products.

The reduction of carbon tetrachloride in aprotic solvents can give dichlorocarbene as intermediate.[8] The formation of 1,1-dichlorocarbene has been demonstrated using lead cathodes in chloroform or dichloromethane media with tetrabutylammonium bromide as electrolyte[9]:

$$CCl_4 \xrightarrow{2e} \bar{C}Cl_3 + Cl^- \longrightarrow \ddot{C}Cl_2 + Cl^-$$

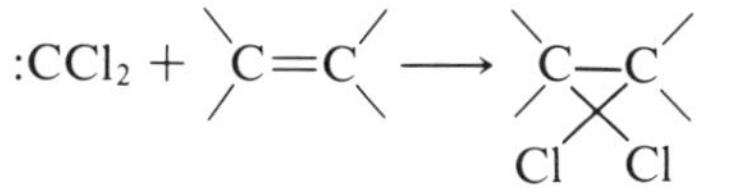

In general, aliphatic halides are reduced at lower potentials than vinyl and aromatic halides, as would normally be expected by analogy with chemical reductions. Stereoisomers are possible with vinyl halides[10]:

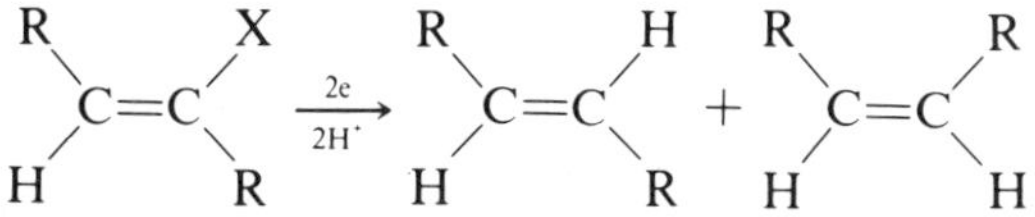

In aprotic media the primary electrodic products cannot be hydrogenated. If they are stable radical ions, they may form organic salts, but usually they are not stable and enter into intermolecular and intramolecular chemical reactions. Rifi, for example, demonstrated various cyclizations in dimethylformamide medium[11]:

$$Br(CH_2)_3Br \xrightarrow{2e} cyclopropane$$

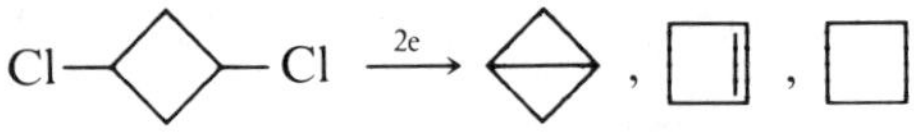

Cipris[12] achieved intermolecular couplings, such as

$$BrCH_2CH_2OH \xrightarrow{2e} \bar{C}H_2CH_2OH + Br^-$$

$$BrCH_2CH_2O^- + CH_3CH_2OH$$

$$\downarrow BrCH_2CH_2OH$$

$$BrCH_2CH_2OCH_2CH_2OH + Br^-$$

Reduction of organic halides is practiced on both laboratory and industrial scale. Various reducing agents are used. (Although dissolving metal reductions, such as the zinc/HCl reduction method, are actually electrochemical processes, we usually prefer to view them as conventional chemical reactions.) The electrolytic method avoids the use of large amounts of reductant and the waste disposal problem may be minor or almost nonexistent. Pentachloropyridine, for example, can be reduced to symmetrical tetrachloropyridine in a sulfonane-water-sodium acetate medium at a lead cathode in 80 to 90% yields.[13] In a solvent mixture of water-tetrahydrofuran-toluene with sodium acetate as electrolyte and at a spongy silver cathode, formed *in situ* on a lead substrate by addition of some $AgNO_3$, in a divided cell, a 1–1 mixture of symmetrical tetrachloropyridine and 1,3,5-trichloropyridine is obtained.[14]

In aprotic media electroreduction of polyhalopyridines gives coupled products[15]:

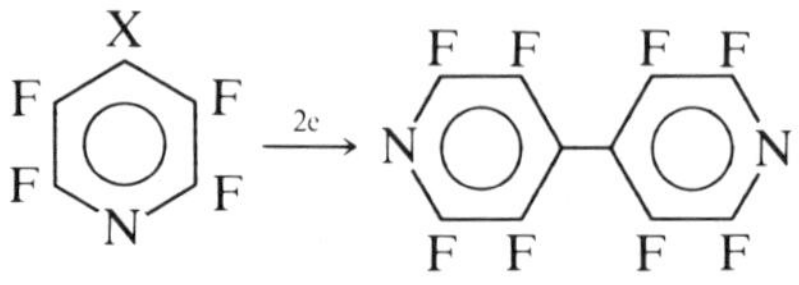

Usually the electrolytic reduction of carbon-halogen bonds is a two-electron reduction. Under certain conditions, however, one-electron reductions occur.

At lead cathodes in acetonitrile solvent, alkyl halides react according to the following mechanism[16]:

$$4CH_3Br + 4e + Pb \longrightarrow Pb(CH_3)_4 + 4Br^-$$

In this case the lead cathode is also a reactant in the usual sense. As we would reasonably expect, catalytic effects by certain positive cations would be possible in carbon-halogen bond electronations. The cation effect may be inferred by analogy to a concerted type action:

$$\overset{\displaystyle\cdot}{\underset{\diagup}{\diagdown}}CBr \cdots Li^+ \longrightarrow \underset{\diagup}{\diagdown}C\cdot + [Br^-Li^+]^0 \text{ iron pair}$$

Such effects can be observed voltammetrically as shifts in the half-wave potentials with various cations. These catalytic effects would be more pronounced in organic solvents of low dielectric constant because in such media ion pairing is most likely to occur.

It would be instructive here to consider the possible pathways in the electroreduction of carbon-halogen bonds. The initial electron transfer, which would most likely be the slowest step in the overall reaction, would be heterogeneous and explained as:

$$\textbf{1} \quad RX + e \rightleftharpoons RX^-$$
$$\textbf{2} \quad RX^- \longrightarrow R\cdot + X^-, 2R\cdot \longrightarrow RR$$
$$\textbf{3} \quad R\cdot + e \xrightarrow[\text{rapid}]{} R^-$$
$$\textbf{4} \quad RX^- + R\cdot \longrightarrow RX + R^-$$

The anion, R^-, is then free to enter into reactions with the medium:

$$R^- + H_2O \longrightarrow RH + OH^-$$
$$R^- + CO_2 \longrightarrow RCOO^-$$

Steps 2 and 4 may take place either at the surface of the electrode or homogeneously in the bulk solution, depending on how stable the species are and how strongly they are adsorbed on the electrode (for a pertinent study see ref. 17).

References

1 M. Von Stackelberg and W. Stracke, *Z. Electrochem.*, **53**, 118 (1949); P. J. Elving, *Record Chem. Prog.*, **14**, 99 (1953); B. J. Casanova and L. Eberson, S. Patai, Ed., in *Chemistry of the Carbon-Halogen Bond*, Interscience, New York, 1974, p. 15.

2 G. M. McNamee et al. *J. Am. Chem. Soc.*, **99**, 1831 (1977).

3 D. Kyriacou, unpublished work.

4 S. Wawzonek and R. C. Duty, *J. Electrochem. Soc.*, **108**, 1135 (1961); S. Wawzonek and J. H. Wagenknecht, *J. Electrochem. Soc.*, **110**, 420 (1963).

5 F. H. Covitz, *J. Am. Chem. Soc.*, **89**, 540 (1967); H. Gilch, *J. Polymer Sci.*, **4**, 1351 (1966).

6 P. J. Elving and B. Pullman, *Advan. Chem. Phys.*, **3**, 1 (1961).

7 S. Wawzonek and J. H. Wagenknecht, *J. Electrochem. Soc.*, **110**, 420 (1963).

8 S. Wawzonek and R. C. Duty, *J. Electrochem. Soc.*, **108**, 1135 (1961).

9 H. P. Fritz and W. Kornrumpf, *Liebigs Ann. Chem.*, **1978**, 1416.

10 A. J. Fry and M. A. Mitnick, *J. Am. Chem. Soc.*, **91**, 6207 (1969).

11 M. R. Rifi, *J. Am. Chem. Soc.*, **89**, 4442 (1967).

12 D. Cripris, *J. Appl. Electrochem.*, **8**, 537 (1978).

13 D. Kyriacou, unpublished work.

14 D. Kyriacou, patent pending.

15 R. D. Chambers, D. I. Clark, C. R. Sargent, and F. G. Drakesmith, *Tetrahedron Lett.*, **1979** (21), 1917.

16 H. E. Uley, *J. Electrochem. Soc.*, **116**, 1201 (1969).

17 C. P. Andrieux, C. Blocman, J. M. D. Buchiat, and J. M. Saveant, *J. Am. Chem. Soc.*, **101**, 3439 (1979).

2.17 Electrocarboxylations

Preparations of organic acids by generation of organic anions by the electrolytic method may prove to be of great synthetic value. It is possible, and easy, to generate organic anions in aprotic media. In the presence of CO_2, which is a good electrophile, formation of carboxylic acids is possible. Even carbon dioxide alone has been reported to give malic acid at a mercury cathode.[1] Stilbene gave the meso-2, 3-diphenylsuccinic acid at a mercury cathode when the electrolysis was carried out in dimethylformamide in the presence of carbon dioxide and tetrabutylammonium iodide as electrolyte.[2] The electrolysis of carbon dioxide to produce oxalic acid and other acids has been pointed out.[3] The electrochemical reduction of benzyl chloride[4] in the presence of carbon dioxide in dimethylformamide gave mandelic acid, benzyl mandalate, and benzyl phenylacetate. The possiblity of electrocarboxylation of organic halides, in general, is referred to in the previous section. In order to minimize protonation of the organic anion, media of low proton availability must be used. Wawzonek and Shradel[4] indicated the formation of lactone by cyclization of the carboxylated intermediate and its subsequent hydrolysis to the mandelic acid.

The preparation of derivatives of 0-carboxymethyltartonic acid has been reported.[5] Electrogenerated bases are produced that deprotonate the organic substrate in the presence of carbon dioxide. The electrochemical reaction is quite complex. The overall cell reaction is

$$+ 2CO_2 + 2Na_2CO_3 \longrightarrow \qquad + 2NaHCO_3$$

The prospects for the electrochemical reduction of carbon dioxide to methanol have been considered.[6] The CO_2 is first reduced to formate, which is then further reduced to methanol in a methanol-HCl medium. The electrolytic carboxylation of acetonitrile and α-substituted acetonitriles has been patented[7]:

Halogenated aromatic nuclei, when reduced in the presence of CO_2, tend to form carboxylic acids[8]:

References

1   A. Bewick and G. P. Greener, *Tetrahedron Lett.*, **1969**, 4623.

2   A. P. Tomilov and B. L. Klyuev, *Zh. Obshch. Khim.*, **39**, 470 (1969) (English ed. p. 446).

3   *Chem. Week*, Jan. 5, (1979).

4   S. Wawzonek and M. J. Shradel, *J. Electrochem. Soc.*, **126**, 401 (1979).

5   C. R. Hallcher, D. A. White, and M. M. Baizer, *J. Electrochem. Soc.*, **126**, 404 (1979).

6   P. G. Russel, N. Kovac, S. Srivinasan, and M. Steinberg, *J. Electrochem. Soc.*, **124**, 1329 (1977).

7   D. A. Tyssee, U.S. Patent 3, 945, 896 (1976).

8   M. I. Rifi, N. L. Weinberg, Ed., in *Technique of Electroorganic Synthesis*, Part II, Wiley-Interscience, New York, 1974, p. 175.

2.18   *N*-Heterocyclic Compounds

Anodic synthesis with *N*-heterocyclic compounds is considered in a previous section 2.8. We indicate briefly the possibilities of the analogous cathodic reaction. For such reactions aqueous sulfuric acid or aqueous-organic media with suitable electrolytes can be used. Depending on the electrode material and the electrolysis medium, various reactions can be accomplished. As an example the reduction of phthalimide derivatives has given various products, selectively[1]:

Vinylpyridines, being activated olefins, can undergo hydrodimerizations[2]:

Electrochemical reduction of pyridopyrazines in aqueous-organic media has afforded 1,4-dihydro derivatives[3]:

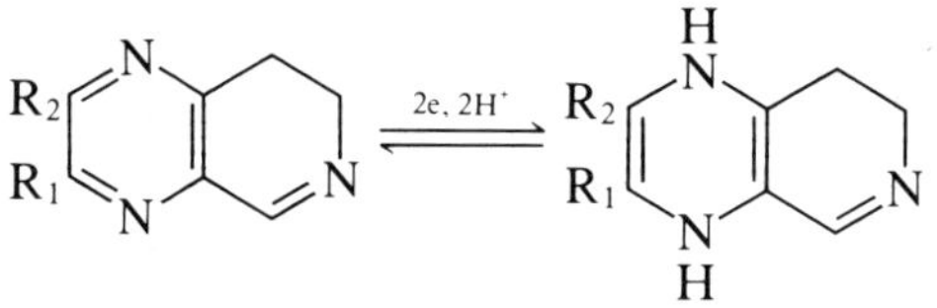

Purine and pyrimidine can be easily reduced at mercury and lead cathodes,[4,5] yielding products such as

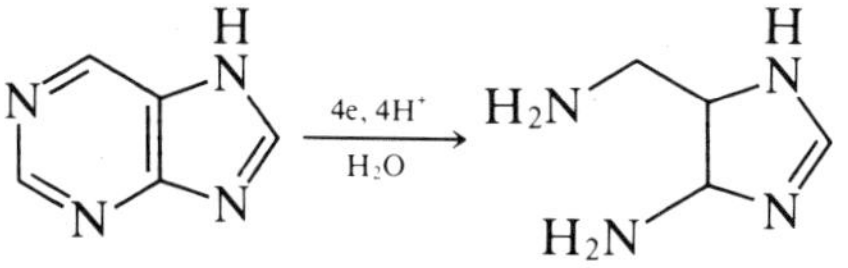

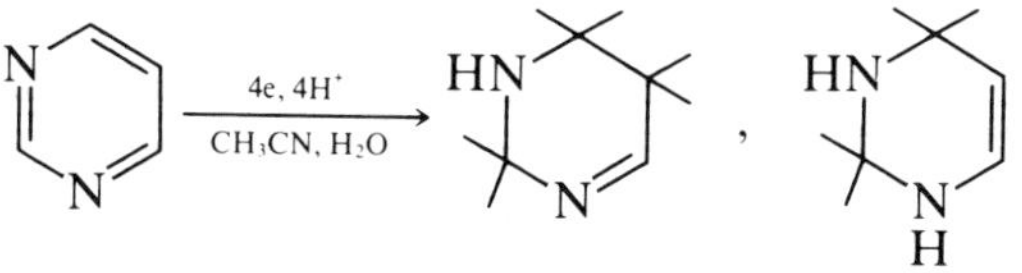

Polarographic studies suggest several synthetic possibilities by reduction of N-heterocyclic compounds: 3,6-diphenyl-2,3,4,5-tetrahydropyrazine has been reduced in acid solution by opening of the nitrogen-nitrogen bond[6]:

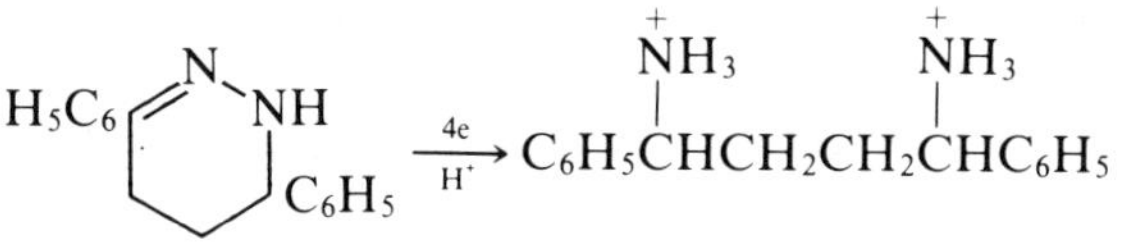

Phthalazinones have a carbon-nitrogen bond reduced before reduction of the nitrogen-nitrogen bond, giving finally an amine and an amide:

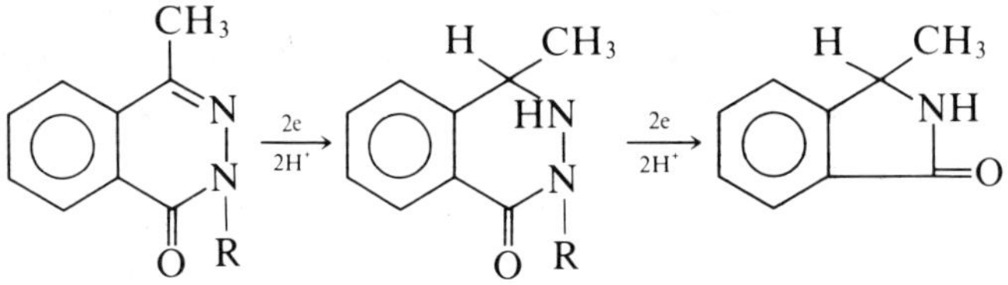

Benzotriazoles are reduced in both acid and alkaline media. Many of these reductions can also be accomplished by chemical methods. The electrolytic method is of value insofar as it can be made selective by using the proper electrode potential, which can be easily determined by polarography. Because many of these substances are sufficiently soluble in aqueous media, their electrolytic reactions should be very attractive.

References

1   B. Sakurai, *Bull. Chem. Soc. Jap.,* **7**, 155 (1932).

2   J. D. Anderson, M. M. Baizer, and E. J. Prill, *J. Org. Chem.,* **30**, 1645 (1965).

3   J. Armand, K. Chekir, and J. Pinson, *Can. J. Chem.,* **56**, 1894 (1978).

4   D. L. Smith and P. J. Elving, *J. Am. Chem. Soc.,* **84**, 2741 (1962).

5   J. E. O'Reilley and P. J. Elving, *J. Am. Chem. Soc.,* **93**, 1871 (1971).

6   H. Lund, *Discuss. Faraday Soc.,* **45**, 193 (1968).

INTRAMOLECULAR ANODIC AND CATHODIC BOND FORMATIONS

Intramolecular bond formations are of great practical importance in several areas of organic synthesis, especially in the area of pharmaceutical synthesis. In the pharmacological research and production fields, the electrolytic method of synthesis may be called upon to supplement or supplant other methods and to open new avenues for syntheses. The large and complex organic and organometallic molecules of biochemical interest could, in many instances, be more specifically modified chemically or stereochemically, and intra- and intermolecularly, at the *inherently disciplined* electrode–solution interface than in the *haphazard* phase of a homogeneous medium. Some examples of intramolecular cyclizations are cited in this section in order to illustrate the potential synthetic value of the electrolytic method.

In most electrolytic reactions of this type, controlled electrode potentials would be required. Selective functional group redox reactions and stereoisomers that were difficult or impossible by conventional methods have been attained by the electrochemical method. To digress a little, the electrochemical method can also be very useful when microquantities of certain substances are needed for initial research studies. *Polarographic synthesis,* long ago recommended (Lingane), has not been appreciated as yet by the organic bench chemist, although as far as microsynthesis is concerned there exists no parallel to the aforementioned method.

An example of stereochemical synthesis and intramolecular bond formation is the following[1]:

The ratio of *cis*- and *trans*-1,2-acenaphthenediols has been found to depend on the kind of acyl group, supporting electrolyte, solvent, and cathode potential.

The cathodic cyclization of iodobenzylisoquinolinium salts has been explained as follows[2]:

Because iodine is the easiest leaving group of all halogens, followed by bromine, chlorine, and fluorine, in that order, intramolecular bonding, such as that shown above, should be attempted via the iodo compounds. Acetonitrile-lithium tetrafluoroborate media are apparently very suitable for intramolecular anodic couplings, such as this one[3]:

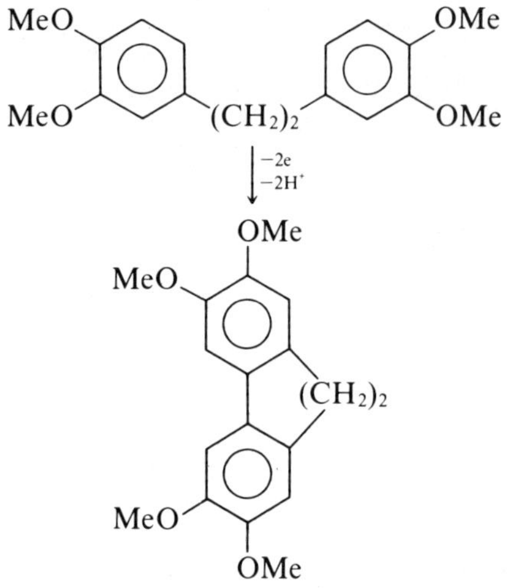

$$
\begin{array}{c}
\text{MeO} \\
\text{MeO} \quad \bigcirc \quad (CH_2)_2 \quad \bigcirc \quad \text{OMe} \\
\text{OMe}
\end{array}
$$

$$\downarrow\; \substack{-2e \\ -2H^{\cdot}}$$

$$
\begin{array}{c}
\text{OMe} \\
\text{MeO} \quad \bigcirc \\
\qquad\qquad (CH_2)_2 \\
\text{MeO} \quad \bigcirc \\
\text{OMe}
\end{array}
$$

Intramolecular cyclization of enol acetates has been accomplished by anodic oxidation in acetic acid/tetraethylammonium p-toluene sulfonate medium[4]:

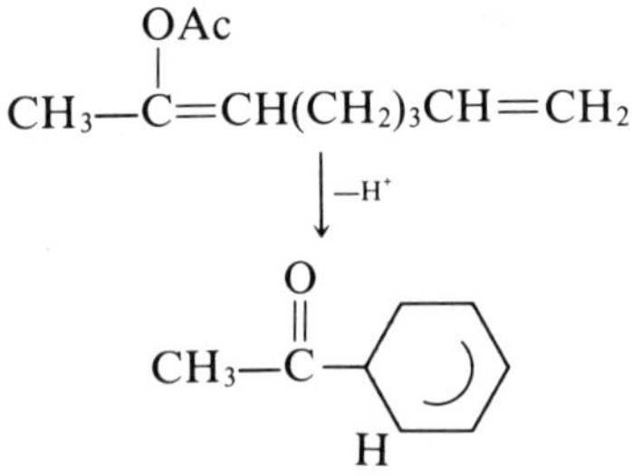

$$
\begin{array}{c}
\text{OAc} \\
| \\
CH_3-C=CH(CH_2)_3CH=CH_2
\end{array}
$$

$$\downarrow\; {-H^{\cdot}}$$

$$
\begin{array}{c}
O \\
\parallel \\
CH_3-C-\bigcirc \\
H
\end{array}
$$

A very easy reductive intramolecular carbon-nitrogen bond formation is the following example[5]:

80%

The reduction takes place at tin or mercury cathodes in a divided cell.

References

1 T. Nonaka and M. Asai, *Bull. Chem. Soc. Jap.*, **51**, 2976 (1978).

2 R. Gottlieb, *J. Am. Chem. Soc.*, **98**, 1708 (1976).

3 V. Pamquist, A. Nilsson, T. Petterson, A. Ronlan, and V. D. Parker, *J. Am. Chem. Soc.*, **44**, 196 (1979).

4 T. Shono, I. Nishicuchi, S. Kashimura, and M. Okawa, *Bull. Chem. Soc. Jap.*, **51**, 2181 (1978).

5 D. Kyriacou, unpublished work.

ELECTROCATALYTIC REACTIONS

In a broad sense all electrochemical reactions are catalytic reactions, since the primary step, the electron exchange, is a heterogeneous catalytic phenomenon; the electrode plays the role of the catalyst, or, more precisely, the *electrocatalyst*, since it catalyzes the electron transfer reaction. Grubb called the phenomenon *electrocatalysis*.[1] The evolution of hydrogen from a platinum electrode is an electrocatalytic reaction. The same reaction occurs at lead and mercury electrodes (high hydrogen overvoltage electrodes). The platinum is apparently a better electrocatalyst for the electroreduction of water than lead or mercury. Nickel, on the other hand, is a better electrocatalyst than the platinum for oxygen evolution from water.

Heterogeneous electrocatalysis involves adsorption of the organic substrate on the electrode surface. The kind of electrode surface and degree of adsorption on various electrode surfaces determines to a large extent the electrocatalytic character of the electrode for a particular reaction. Many organic electrode reactions are possible at one electrode and not at another (or more efficient at one than at another), all other conditions being the same. Such differences arise from differences in electrocatalytic properties of the electrodes. An interesting example is included in Chapter 4.

There are also many possibilities for homogeneous electrocatalysis. In such cases the generation and regeneration of the catalyst is actually a heterogeneous electrochemical reaction (which in itself might be an electrocatalytic reaction involving the electrode as a catalyst), but the chemical reactions in the bulk of the solution are catalyzed by the electrogenerated catalyst that goes into solution, and are similar to conventional homogeneous catalytic reactions. Schematically, homogeneous electrocatalysis can be shown as follows:

$$Ox + e \longrightarrow Red \qquad catalyst$$

$$Red + S \longrightarrow product + Ox$$

$$Ox + e \longrightarrow Red \qquad regeneration$$

Generation of oxidants capable of acting as catalysts is also possible. For example, the oxidation of alcohols by iodonium ion can be considered a catalytic reaction, the iodonium ion being the catalytic electron transfer agent[2]:

$$I^- \longrightarrow I^+ + 2e \qquad at\ anode$$

$$I^+ + RCH_2OH \xrightarrow{-H^-} RCH_2OI \longrightarrow RCHO + I^- + H^+$$

The aldehyde may then be further oxidized, either by I^+ or directly at the anode, to give products that depend on the electrolysis medium. In principle only catalytic amounts of iodide ion suffice. Known oxidizing or reducing systems can be used in catalytic amounts in electrolytic synthesis. Many examples can be found in the literature. Chromium (II) has been used in dimethylformamide for the reductive coupling of benzylic and allylic halides.[3] The known $Fe(II)/H_2O_2$ reagent has been used for the oxidation of polymethylated benzenes.[4] Catalytic amounts of ferrous ion can be used. The ferric-ferrous redox couple is reversible so that the formation of the actual oxidizing species, namely the $OH^\cdot$ radical, can be continuous:

$$Fe^{3+} + e \longrightarrow Fe^{2+} \qquad at\ cathode$$

$$O_2 + 2e + H^+ \longrightarrow H_2O_2 \xrightarrow{Fe^{2+}} OH^\cdot + OH^-$$

Electrosyntheses can often be accomplished in one-compartment cells where the cathodic and anodic products react as they are formed to give new products useful in themselves or as intermediates. Consider the following generalized possibility:

$$R + e \longrightarrow R \bar{\vee} \quad \text{at the cathode}$$

$$X - e \longrightarrow X^+ \vee \quad \text{at the anode}$$

$$R\bar{\vee} + X^+\vee \longrightarrow RX \quad \text{in solution}$$

$$RX + AB \longrightarrow RA + B + X \quad \text{in solution again}$$

Substance X can be a catalyst continuously regenerated in the cell. Thus electrosynthetic hetero-hetero atom bond formations have been promoted by sodium bromide.[5] The electrocatalytic preparation of sulfinimides from phthalimide and dicyclohexyl disulfide in acetonitrile containing a catalytic amount of sodium bromide has been shown to be feasible (undivided cell):

$$2R'(CO)_2NH \xrightarrow{2e} 2R'(CO)_2N^- + H_2$$

$$\diagup \text{RSSR}$$

$$\diagup 2Br^- \xrightarrow[\text{anode}]{-2e} Br_2$$

$$R'(CO)_2NSR + RSBr$$

The same principle has been applied to the synthesis of phosphorothiolates:

$$(RO)_2P(O)H + R'SSR'$$

$$\Big\downarrow \begin{smallmatrix} -e \\ NaBr \end{smallmatrix}$$

$$(RO)_2P(O)SR'$$

An interesting example of electrocatalytic oxidation is that of o-nitrotoluene to o-nitrobenzaldehyde with electrogenerated cobaltic ion.[6] The cobaltic ion oxidizes the organic molecule and the resulting cobaltous ion is reoxidized at the anode to cobaltic ion to repeat the cycle.

Selective cyclodimerization of 1,3-diolefins has been possible, at room temperature, by electrocatalysis with $Fe(NO)_2$ as catalyst. This catalyst was prepared *in situ* by electrolytic reduction of $FeCl_3$ in the presence of NO[7]:

$$[Fe(NO)_2Cl]_2 + 2e \longrightarrow 2Cl^- + 2Fe(NO)_2$$

All these homogeneous electrocatalytic reactions are indirect oxidations or reductions of organic compounds. They are especially desirable because only catalytic amounts of the redox catalysts may be needed, and they can be faster than heterogeneous electrodic reactions. Although in many cases the catalytic activity may be related to the standard redox potential of the couple, it must be remembered that such inference cannot always be valid. This is because in many cases it would be the kinetics, rather than the thermodynamics of the redox system, that determine catalytic effects. It is obvious that, although homogeneous electrocatalysis may have technical and economic value, the selectivity that is inherent in many heterogeneous electrochemical reactions is almost lost. Nonetheless research in this field of electrocatalysis for synthetic purposes holds very great promise. Active, and possibly selective, transient species may be electrogenerated *in situ* or *ex situ,* thus avoiding exotic and expensive reagents that nonelectrochemical methods so often require.[8]

References

1   T. W. Grubb, *Nature,* **198,** 883 (1963).

2   T. Shoho, Y. Matsumura, J. Hayashi, and M. Mizguchi, *Tetrahedron Lett.,* **1979,** 165.

3   J. Wellmann and E. Steckhan, *Synthesis,* **1978** (12), 901.

4   R. Tomat and A. Rigo, *J. Appl. Electrochem.,* **9,** 301 (1979).

5   S. Torii, H. Tanaka, and V. Ukida, *J. Org. Chem.,* **44,** 1554 (1979); S. Torii, H. Tanaka, and N. Sayo, *J. Org. Chem.,* p. 2938.

6   C. Comninellis, E. Plattner, and P. Jaret, *J. Appl. Electrochem.,* **9,** 753 (1979).

7   D. Huchette, J. Nicole, and F. Petit, *Tetrahedron Lett.,* **1979** (12), 1035.

8   X. D. Hemptinne and K. Scunk, *Trans. Faraday Soc.,* **65,** 591 (1969); V. I. Gants, E. A. Smirnova, V. V. Sysoeva, and N. N. Storchak, *J. Appl. Chem.* (USSR), **42,** 2345 (1969); J. P. Collman, *Accounts Chem. Res.,* **1,** 136 (1968); M. M. Baizer and D. Carter, U.S. Patent 3,545,083 (1970); J. W. Schultze and M. A. Habib, *J. Appl. Electrochem.,* **9,** 255 (1979); C. P. Andrieux, C. Blocman, J.-M. D. Bouchiat, and J.-M. Saveant, *J. Am. Chem. Soc.,* **101,** 3439 (1979); J. Pinson and J.-M. Saveant, *J. Am. Chem. Soc.,* **100,** 1506 (1978); H. Lund and L. H. Kristensen, *Acta Chem. Scand.,* **B33,** 495 (1979).

3
SOME SPECIAL TOPICS

3.1 Amalgams

Amalgams are alloys of mercury with various metals or nonmetals. Reduction of organic compounds with amalgams is a very old and widely practiced technique. Amalgams can be formed *in situ* or *ex situ* by electrolyzing solutions of metal salts using a pool of mercury as cathode. Many types of organic compounds can be reduced with amalgams: carbonyls, halides, olefins, nitro compounds, and various molecules with reduction potentials up to about —2 V or higher, depending on the type of amalgam. In symbols the reduction is shown as

$$M(Hg) + Ox \longrightarrow Red + M^{n+} + Hg$$

The amalgam, or better, the amalgamated metal, M, acts as an electron donor in the same sense that a cathode in an electrolytic cell does. However, in the latter case, the potential of the electrode, and hence its reducing ability, can be varied, whereas the potential of the amalgam is usually fixed. At best it varies somewhat with the concentration and kind of metal that is dissolved in the mercury. In addition to metal amalgams, ammonium ion and organic amalgams are also formed, namely alkylammonium, sulfonium, and phosphonium amalgams. Organic amalgams decompose upon warming to room temperatures[1]:

$$(CH_3)_4 \overset{\cdot}{N}(Hg) \longrightarrow CH_3N + CH_3^- + Hg$$

Space does not permit further discussion of this subject. Comprehensive discussions can be found in refs. 2 and 3.

References

1   H. N. McCoy and W. C. Moore, *J. Am. Chem. Soc.*, **33**, 273 (1911).

2   H. Lund, in M. M. Baizer, Ed., *Electroorganic Chemistry*, Marcel Dekker, New York, 1973, p. 805.

3   W. J. Settineri and L. D. McKeever, in N. L. Weinberg, Ed., *Technique of Electroorganic Synthesis*, Part II, Wiley-Interscience, New York, 1975, p. 397.

3.2   Organometallics

The field of organometallics is an important specialty in itself and it is impossible to even sketch it in this book. Suffice it to state here that many organometallic compounds can be synthesized by the electrolytic process. The manufacture of tetraethyl lead is well known.[1] At the surface of the lead anode alkyl radicals are formed by anodic oxidation and abstract atoms of lead to form tetraethyl lead. This type of reaction is typical of many electrochemical reactions where the electrode itself reacts with the organic substrate.

Several examples of electroorganometallic preparations are described in the literature.[1-5] Tetraalkyl derivatives of tin and lead using alkyl sulfates have been obtained in yields between 7 and 84% using zinc cathodes and lead or tin as anodes.[2] The presence of catalytic amounts of alkyl iodides strongly aids the formation of these organometallic products.

The direct synthesis of organometallic compounds using the so-called sacrificial electrodes in nonaqueous media has been reviewed recently.[3] Many organometallic halides and metal-chelates have been prepared by the electrochemical method cathodically and anodically. Such electrochemical reactions can be symbolized as

$$RX + e \longrightarrow [RX]^- \longrightarrow R^{\cdot} + X^-$$

$$\text{metal (M)} + R^{\cdot} \longrightarrow MR$$
$$\text{(cathode)}$$

For example:

$$C_2H_5Br \xrightarrow[\substack{CH_3CN \\ medium}]{Sn} Sn(C_2H_5)_4$$

$$RX \xrightarrow[tin\ anode]{} R_2SnX_2$$

$$CH_3I \xrightarrow[Sn]{-e} (CH_3)_2SnI_2$$

Very active metallic complexes of organic compounds may be formed directly in an electrochemical cell for specific syntheses. It has been shown that nickel and iron acetylacetonates can be electrochemically formed in a cell for the synthesis of hexadecane from 1-octyl bromide.[6] The metallic complexes acted as catalysts for this reaction, since in their absence little or no hexadecane was formed.

References

1   Nalco Chemical Co., U.S. Patent 3,256,161 (1961).

2   G. Mengoli, S. Daolio, and F. Furlanetto, *Ann. Chim.* (Rome), **68**, 455 (1978).

3   D. G. Tuck, *Pure Appl. Chem.*, **51**, 2005 (1979).

4   H. Lehmkul, in M. M. Baizer, Ed., *Organic Electrochemistry*, Marcel Dekker, New York, 1973, p. 621.

5   W. J. Settineri and L. D. McKeever, in N. L. Weinberg, Ed., *Technique of Electroorganic Synthesis*, Part II, Wiley-Interscience, New York, 1975, p. 397.

6   P. W. Jennings, D. G. Pillsbury, J. L. Hall, and V. T. Brice, *J. Org. Chem.*, **41**, 719 (1976).

3.3   The Solvated Electron

We have already formally recognized the electron as a transient reactant. We now consider briefly the existence of the *solvated* electron in the bulk of certain media, such as liquid ammonia, amines, and ethers.[1]

If a cathode is made very negative ($E^0 = 2.68$ V for hydrated electrons), electrons will be ejected from the cathode into the surrounding medium, as if they were negative ions, and they may become solvated by the molecules of the medium.[2] Such electrolytically generated *free* electrons are extremely active species and can *electronate* most substances. Reduction by free electrons is a homogeneous reaction and is usually nonselective because of the very great reactivity of the electrons. In Dainton's view, the electron is the simplest and most powerful free radical.[6] The primary reaction can be expressed as follows:

$$e \quad + \text{solvent} \longrightarrow \quad e$$
$$\text{(cathode)} \qquad\qquad\qquad \text{(solvent)}$$

Following this primary event reaction with the solvent itself, or with a substance in the solvent, occurs:

$$NH_3 + H_2O \longrightarrow NH_4^+ + OH^-$$

$$e(NH_3) + NH_4^+ \longrightarrow 2NH_3 + \tfrac{1}{2}H_2$$

$$e(NH_3) + A \longrightarrow A^{\cdot} + NH_3$$

Using a low molecular weight amine, for instance, various substances can be reduced by catalytic generation of electrons in the presence of lithium chloride. The applied potential must be negative enough to cause reduction and plating out of lithium on the electrode surface. The electrons are then freed into the solution by the heterogeneous reaction of metallic lithium with the medium:

$$Li^+ + e \longrightarrow Li^0 \; (Pt)$$

$$Li^0 + RNH_2 \longrightarrow Li^+ + e(RNH_2)$$

In principle, only catalytic amounts of lithium would be required. Aromatic and aliphatic nitriles have been reduced to their corresponding hydrocarbons by solvated electrons[3,4]:

$$RCN + e \longmapsto RCN^{\cdot}$$
$$\text{(solvent)}$$

$$RCN^{\cdot} \longrightarrow R^{\cdot} + CN^-$$

$$R^{\cdot} + e + H^+ \longrightarrow RH$$

It is possible to use mixed solvents provided one component is capable of solvating electrons. It has been shown that benzene and olefins can be reduced by electrolytically generated electrons in a medium composed of ethanol and hexamethylphosphoramide.[5] The latter solvent is known to solvate electrons. Solutions containing free electrons are blue.

A very illuminating exposition of the electron as a chemical entity has been made by Dainton.[6] Tuttle and Graceffa[7] raised the question, "Solvated Electron or Not?" The reader will find their paper stimulating.

References

1 E. J. Hart and M. Anbar, *The Hydrated Electron*, Wiley-Interscience, New York, 1970; E. J. Hart, *Accounts Chem. Res.*, **2**, 161 (1969); A. J. Birch, *Nature*, **158**, 60 (1946).

2 E. J. Hart, S. Gordon, and E. M. Friedman, *J. Phys. Chem.*, **70**, 150 (1966).

3 R. A. Benkeser and E. M. Kaiser, *J. Am. Chem. Soc.*, **85**, 2858 (1963).

4 P. G. Arapakos, M. K. Scott, and F. E. Huber, Jr., *J. Am. Chem. Soc.*, **91**, 2059 (1969).

5 H. W. Sternberg, R. E. Marby, I. Wender, and D. M. Mohilner, *J. Am. Chem. Soc.*, **91**, 4191 (1969).

6 F. S. Dainton, *Q. Rev. Chem. Soc.*, **1975**, 323.

7 T. R. Tuttle and P. Graceffa, *J. Phys. Chem.*, **75**, 843 (1971).

3.4 Electrogenerative Organic Systems

Inorganic spontaneous electrochemical systems are many, as for example, the zinc-copper cell and all usual batteries. In all such *spontaneous* systems current is generated as a result of the chemical reactions. These systems represent rapid heterogeneous chemical reactions where the free energy is negative. It is sometimes possible to select organic reactions or systems with negative free energies that are therefore potentially capable of acting like (organic) batteries, provided the reactions are rapid. If a suitable catalytic electrode surface is used, organic reactions can become rapid and able to generate electrical energy.

Organic spontaneous electrocatalytic systems have been called *electrogenerative* to indicate their ability to produce current. One such typical system is the chlorine-ethylene cell[1]:

$$Cl_2 + 2e \longrightarrow 2Cl^- \qquad \text{at the cathode}$$

$$C_2H_4 - 2e \xrightarrow[Cl^-]{} CH_2ClCH_2Cl \qquad \text{at the anode}$$

The electrode over which chlorine is passed adsorbs chlorine molecules. The adsorbed chlorine abstracts electrons from the electrode and it is converted to chloride ions. At the anode the ethylene is deelectronated and it then reacts with chloride ions to form 1,2-dichloroethylene.

Another example is the hydrogenation of benzene.[2] At the anode hydrogen is oxidized to protons, while at the cathode benzene is hydrogenated:

$$3H_2 \longrightarrow 6H^+ + 6e \qquad \text{at the anode}$$

$$6H^+ + 6e \xrightarrow[C_6H_6]{} C_6H_{12} \qquad \text{at the cathode}$$

References

1 S. J. Pietsch and S. H. Langer, in M. Krumptelt, E. Y. Weissman, and R. C. Alkire, Eds., *Electroorganic Synthesis Technology*, Vol. 75, AICE, New York, 1979, p. 51.

2 S. H. Langer and S. Yurchak, *J. Electrochem. Soc.*, **9**, 1228 (1969).

3.5 Electrolytic Formation of the Superoxide Ion and of Ozone

In aqueous media oxygen is reduced very easily to hydrogen peroxide or water, depending on pH. In aprotic media containing quaternary ammonium salts as electrolytes, oxygen is reduced to superoxide ion,[1] which is a fairly stable species at room temperatures. As would be expected it can be a strong nucleophilic species[2]:

$$O_2 + e \longrightarrow O_2^-$$

$$RX + O_2^- \longrightarrow RO_2^- + X^-$$

$$RO_2^- + O_2^- \longrightarrow RO_2^- + O_2$$

$$RO_2^- + RX \longrightarrow ROOR + X^-$$

Synthetic applications have not yet been proven on a significant scale, but because electrolytic formation of superoxide ion is quite easy, its synthetic potential is not to be overlooked.

It would be of interest to mention here the formation of ozone by electrolysis of aqueous solutions at room temperatures. At lead dioxide anodes in aqueous phosphate media, a mixture of oxygen and ozone gases is evolved at the anode.[3] Even though this electrosynthesis is of the inorganic type, it may find organic applications.

References

1 M. Peover and B. S. White, *Electrochim. Acta.*, **11**, 1061 (1966).

2 A. L. Berre and Y. Berger, *Bull. Soc. Chim., Fr.,* **1966**, 2363; M. V. Merritt and D. T. Sawyer, *J. Org. Chem.*, **35**, 2157 (1970).

3 H. Fritz, J. C. G. Thanos, and D. W. Wabner, *Z. Naturforsch.*, **34b**, 1617 (1979).

3.6 The Molecule in the Electric Field

Although science continues to replace art in electroorganic chemistry, much of the rigorous science applies only to ideal systems.

We recall that an organic molecule may find itself immersed in an electric field of 10^7 V/cm. In such an enormous field the molecule may suffer serious deformations and may even have some bonds severely distorted or completely severed because of the energies imparted to these bonds from the field. Adsorption and orientation effects at the electrode surface will be influenced by the field in various unpredictable ways, and these effects alone may sometimes override other structural effects associated with the *normal* physicochemical properties of the molecule. In such a situation theoretical models based on considerations of the single molecule under zero field conditions might be of little or no practical value. It might not be so bold to assume that sometimes the aromatic nature of an adsorbed organic molecule is destroyed as a result of electrostatic forces alone. Some electroorganic reactions have been explained on grounds of field effects. The cathodic reduction of benzil, for example, was found to give a mixture of *cis-* and *trans-*stilbene diols in aqueous solutions[1]:

$$
C_6H_5COCOC_6H_5 \xrightarrow{2e}
\begin{array}{c} C_6H_5 \quad C_6H_5 \\ | \quad\quad / \\ C = C \\ | \quad | \\ OH \ OH \end{array}
+
\begin{array}{c} C_6H_5 \quad OH \\ | \quad\quad / \\ C = C \\ | \quad\quad \backslash \\ OH \quad C_6H_5 \end{array}
$$

The ratio of these isomers depends heavily on the electric field strength (electrode potential). The field was believed to partially destroy the double bond character of the trans isomer, thus permitting rotations about the bond not favoring the formation of the cis isomer.

Adsorbed films, depending on the orientation of the adsorbed organic material and on the orientation of the organic molecule that is to exchange electrons with the electrode, may affect the electron transfer rate and the nature of the products.[2] The reduction of 4-cyanopyridine at a mercury cathode is most probably influenced by field orientation forces[3]:

Although the same cyano group was reduced, the products were different. Apparently, the molecule is oriented in two dissimilar ways at the two potentials. We must also bear in mind that a molecule may not always be able to retain similar electrophoric centers at different potentials. Thus although at one potential a cyano group may be reduced so as to result in carbon-carbon rupture, at a higher potential (electric field) the electrophoric properties of the molecule may be modified so that the molecule is electronated or deelectronated differently. Instead of carbon-carbon rupture, reduction of the cyano group to the amine may occur if the aromatic character of the pyridine ring is destroyed by adsorption forces at the higher potential. Field effects would be almost impossible to predict, especially with complex and large organic molecules.

Field effects have been seriously considered in connection with electrochemical synthesis of polymers.[4]

References

1   A. Vincent-Chodowska and Z. R. Grabowski, *Electrochim. Acta*, **9**, 789 (1964).

2   J. Heyrovsky and J. Kuta, *Principles of Polarography,* Academic, New York, 1966; R. S. Missan, E. I. Becker, and L. Meites, *J. Am. Chem. Soc.,* **83**, 58 (1961).

3   J. Volke and A. M. Kardos, *Collect. Czech. Chem. Commun.*, **33**, 2560 (1968).

4   B. L. Funt and J. Tanner, in N. L. Weinberg, Ed., *Technique of Electroorganic Synthesis,* Part II, Wiley-Interscience, New York, 1975.

4

THE PRAXIS OF ELECTROORGANIC SYNTHESIS

SAFETY PRECAUTIONS IN USING ELECTRICITY

Familiarization with the basic safety practices in the use of electricity must be an imperative in the practice of electrosynthesis. As a basic rule *correct grounding* of the electrochemical unit should precede all other activity. Another basic rule is *never* to touch the wires or electrodes while the power switch is on. These rules should be applied as a matter of routine even though at low voltage (less than 20 V) there is no danger under normal conditions. Always avoid wet hands while working with electricity. Precautions must be taken against spark formation or local overheating caused by faulty electrical connections. During an electrolysis potentially explosive mixtures of gases may be formed (H_2, O_2, Cl_2, etc.). Protective measures must be taken to eliminate the danger of explosions.

Figure 4.1 is a diagram illustrating correctly grounded apparatus. The apparatus should never be assumed to be grounded. The following procedure for grounding is recommended:

1 Before connecting the apparatus to the power source, insure that the electrical resistance of every metallic or conducting part of the equipment to

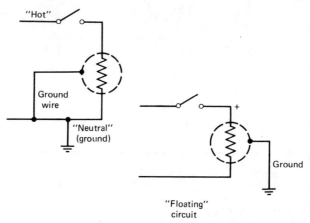

Figure 4.1 Correctly grounded equipment.

the ground is *zero*. Using a suitable device, such as a multimeter, firmly press one of the probes of the measuring device against the electrical ground wire while firmly pressing the other probe against every exposed metallic part of the equipment sequentially. (Sometimes it is necessary to scratch the metallic surface for electrical connection.) The selector switch of the multimeter is set to the lowest range of resistance for these measurements. All readings thus obtained will be *zero resistance* if all metallic parts are electrically connected to one another as they should be.

2 Disconnect the multimeter and connect the equipment to the power source. With the selector switch set to voltage (which would be dc for electrolysis) the testing procedure is repeated as with the resistance measurements. All readings now should indicate 0 V. If both resistance and voltage readings are zero, the equipment is correctly grounded. It should be understood that as low as 75 mA of current through the human body can cause death. The body can usually detect as little as 1 mA as an irritating or tingling sensation. More than a few milliamperes cause shock and muscular immobility. Always check the plug prongs on ac circuits for proper phasing, making certain that they are not reversed. If the prongs are reversed, this should be corrected immediately. For beginners *it would always be advisable to invite an electrician when first installing an electrochemical apparatus.*

This brief reference to safety is not meant to imply that the electrochemical method is intrinsically less safe than the conventional methods. In fact the contrary may be claimed. Seldom, if ever, are very high temperatures and pressures necessary for electrolytic reactions, and the probability of a *runaway* reaction is virtually zero, since the electrolytic reaction may be stopped almost instantly by simply turning off the power switch.

PREPARATIVE EXAMPLES

4.1   Reductive Intramolecular Carbon-Nitrogen Bond Formation

Conversion of 2-phthalidyl-1(2H)-phthalazinone to phthalazino(2,3-b)phtha-lazine-5,12(7H,14H)-dione.

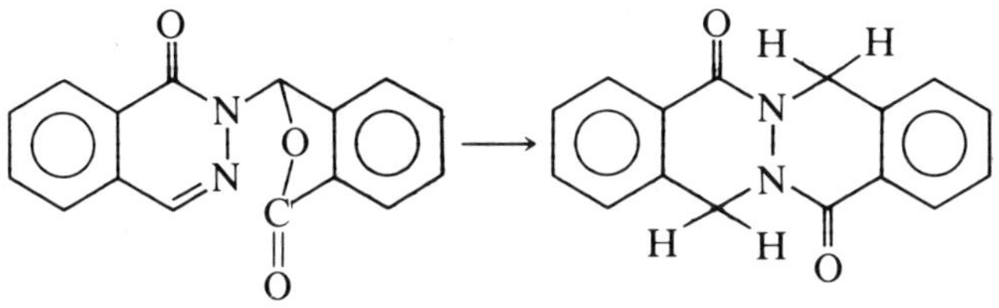

1   **2**

From organic chemistry it can be inferred that, if substance **1** can be reduced, chemically or catalytically, to substance **3**:

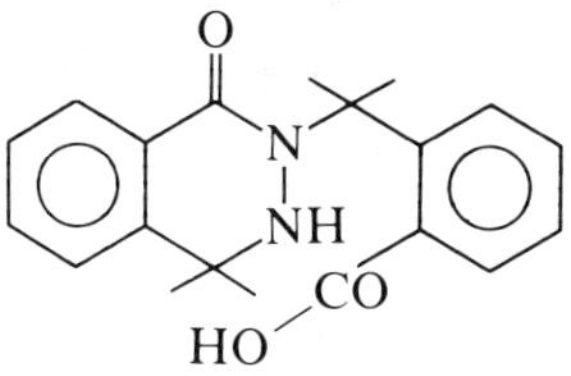

3

the desired compound **2** can be obtained by elimination of a molecule of water from the intermediate **3**. It is reasonable to expect, therefore, that electrolytic reduction of **1** at 90 to 95°C would result directly in the formation of **2**.

   A polarographic wave of the starting material (**1**) can be developed in aqueous media of pH 5.5 to 6.5 containing sodium acetate and some HCl. The polarographic wave is poor, but it shows that reduction occurs at about −1.3 to −1.5 V versus SCE. Also, a voltammogram using tin wire as cathode shows the possibility of the reduction of **1**. It remains to prove by preparative scale

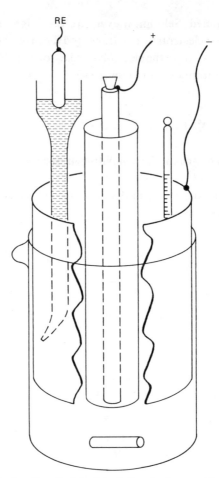

Figure 4.2   Electrolysis cell: cylindrical tin foil cathode, graphite rod anode, ceramic cup for anode compartment. RE: reference electrode.

electrolysis the possibility of the desired synthesis. Because compound **1** is only slightly soluble in water, a suitable aqueous-organic medium must be used for the electrolysis. After several trials it has been found that a 1:1 mixture of water and sulfolane was suitable. Maximum yields of **2** can be obtained in a divided cell, using a tin or mercury cathode, a pH of 5.8 to 6.2, and temperatures of 90 to 95°C. The electrolysis cell is a 250 ml glass beaker at the center of which is positioned a porous ceramic cup to contain the anolyte and the anode. The tin cathode is a tin foil shaped to form a cylinder fitting into the inner wall of the beaker, and the anode is a graphite rod the size of an ordinary pencil, which is placed at the center of the porous cup. The cell assembly is illustrated in Fig. 4.2. Agitation is provided by a magnetic bar, the cell being placed in a mineral oil bath on top of an ordinary hot plate in order to maintain the temperature at 90 to 95°C during the electrolysis. Into the beaker are placed 25 ml of water, 25 ml of sulfolane, 5 g of starting compound 1, and 4 g each of sodium chloride and sodium acetate trihydrate. A few drops of conc. HCl is also added to adjust the pH to 6.0. To the porous cup are added 20 ml of 15% HCl. The cathode (tin) is connected to the negative terminal (working electrode) and the anode (graphite) to the positive terminal of the potentiostat (Princeton Applied Research, Model 371). The reference electrode is a commercial saturated calomel electrode (SCE) situated in the well of the Luggin capillary, as shown in Fig. 4.2. The cathode potential is set to -1.35 V by using the knob in the potentiostat and is automatically kept at that value during the entire run. The power switch is turned on, the solution is stirred, and a current of $\sim$0.5 A begins to flow. As the temperature rises to 90 to 95°C the current rises to $\sim$1.5 A. The pH of the catholyte is kept at 5.8 to 6.2 by periodic additions of drops of conc. HCl. After 5 hr of current passage the electrolysis is stopped. The catholyte is diluted with 500 ml of warm water (60°C), whereupon copious precipitation of the product occurs. The solids are filtered, washed with water, and dried under vacuum at 60°C, giving 3.9 g of 98% of the desired compound 2. This corresponds to 82% yield. Identification of the product is made by comparison with an authentic sample using infrared and mass spectroscopic methods and vapor phase chromatography.

Because hydrogen is also evolved at the cathode in this preparation, it is not easy to determine the number of electrons involved in this cathodic reaction. We can assume that the cathodic reactions are as follows:

$$2H_2O + 2e \longrightarrow H_2 + 2OH^-$$

The anodic reaction is evolution of chlorine, since chloride ion is in large excess relative to OH^- ion in the anolyte solution:

$$4Cl^- \longrightarrow 2Cl_2 + 4e$$

If the reduction of water is neglected, the cell reaction would be written as

$$1 + 4e + 4H_2O \longrightarrow 3 + 4OH^-$$
$$3 \xrightarrow[90-95°]{(-H_2O)} 2 + H_2O$$
$$4Cl^- - 4e \longrightarrow 2Cl_2$$

The overall reaction is

$$1 + 3H_2O + 4Cl^- \longrightarrow 2 + 2Cl_2\uparrow + 4OH^-$$

D. Kyriacou, unpublished work.

4.2   Carboxylic Acids from Primary Alcohols

Oxidations of long chain and certain unsaturated primary alcohols to carboxylic acids are very easily accomplished by the electrolytic method. Conventional chemical oxidations (permanganate, nickel peroxide, chromate) are less convenient and give low yields. Chemical methods may affect the unsaturated bonds, whereas such bonds are unaffected or less affected by the anodic oxidation method.

Using the nickel hydroxide anode, alcohols such as

$$n\text{-}C_nH_{n+1}\text{---}CH_2OH, \quad ArCH_2OH, \quad \langle\!\!\!\langle \quad \rangle\!\!\!\rangle \; CH_2OH ,$$

$$\diagdown\!\diagdown CH_2OH, \quad HC\!\equiv\!CCH_2OH$$

are oxidized to their carboxylic acid analogs in yields of 50 to 90% in aqueous or in aqueous-t-butanol-NaOH media at 25 to 75°C. The preparation of stearic acid from octadecanol is described here as a typical example:

$$n\text{-}C_{17}H_{35}\text{---}CH_2OH \xrightarrow[\text{Ni, 75°C}]{H_2O/NaOH} n\text{-}C_{17}H_{35}\text{---}\overset{\displaystyle O}{\overset{\|}{C}}\text{-}OH$$
$$77\%$$

The electrolysis is carried out in an undivided double-walled 300 ml cylindrical glass cell equipped with a 250 cm^2 nickel net as anode and a stainless steel piece as cathode. The nickel net is converted to the active nickel hydroxide electrode by treatment with a low frequency alternating current in 0.1 N nickel sulfate/0.1 N sodium acetate/0.005 N sodium hydroxide solution.

A mixture of octadecanol (8.12 g, 30 mmol) and 1 M aqueous sodium hydroxide (250 ml) solution is electrolyzed for 8 hr at 75°C using a current of 4 A (16 mA/cm^2) and a cell voltage of 2.0 V. Sodium stearate precipitates, the total sodium stearate product is dissolved by addition of *t*-butanol (20 ml), and the stearic acid is obtained by precipitating it as the barium salt by adding 400 ml of saturated barium hydroxide solution. The barium stearate salt is filtered, washed with water and ether, and dissolved in 15% hydrochloric acid (100 ml). The solution is extracted with ether, (3 × 100 ml), dried with sodium sulfate, and evaporated. The residual product is purified by bulb-to-bulb distillation at 0.01 torr; yield: 6.56 g (77%); m.p. 68 to 69°C.

J. Kaulen and H. J. Schafer
Synthesis, **1979**(7), 513–516
Adapted from *Synthesis*, by permission of the copyright owner.

4.3 Cyanation of *N-n*-Propylpyrrolidine

The electrolysis is performed in a 100 ml divided cell, using a porous cup for the catholyte. The anode is a cylindrical platinum net (4.5 cm in height, 11.0 cm in circumference, 55 mesh) and the cathode is a platinum *coil*. The reference electrode is the SCE. The electrolysis is carried out at 3°C with constant stirring by a magnetic bar. The anolyte contains 0.1 mol of the amine and 0.15 mol of sodium cyanide in 75 ml of 1 : 1 methanol-water solution. The catholyte is an aqueous methanol solution of sodium cyanide. Current is passed for 8 hr at an anode potential in the range of 1.1 to 1.5 V so as to maintain a constant current of 0.5 A during the entire electrolysis (0.15 F). At the conclusion of the electrolysis the anolyte is concentrated under reduced pressure at 35°C on a rotary evaporator. The remaining liquid is saturated with anhydrous potassium carbonate. The organic layer is separated and the aqueous layer is extracted twice with 30 ml portions of ether. The combined extracts are dried with potassium carbonate and then analyzed by GLC. The distillation of the crude product gives 8.2 g of colorless liquid at 114 to 119°C (40 mm), which is almost pure N-n-propyl-2-cyanopyrrolidine as confirmed by IR, NMR, and elemental analysis of the picrate derivative. The yield and current efficiency are 59 and 79%, respectively.

T. Chiba and Y. Takata
J. Org. Chem., **42**, 2973 (1977)
Adapted with permission, copyright 1962, American Chemical Society.

4.4 Preparation of 3,6-Dichloropicolinic Acid from 3,4,5,6-Tetrachloropicolinic Acid

The proposed electroreduction is carried out in strongly alkaline aqueous medium. We may write the desired overall cell reaction:

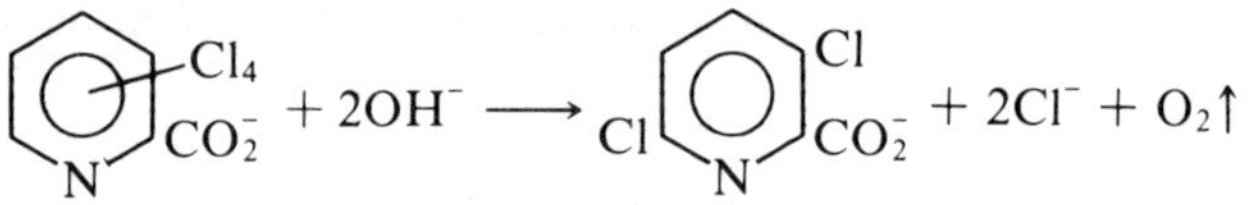

This equation is the sum of the expected cathodic and anodic reactions, which are

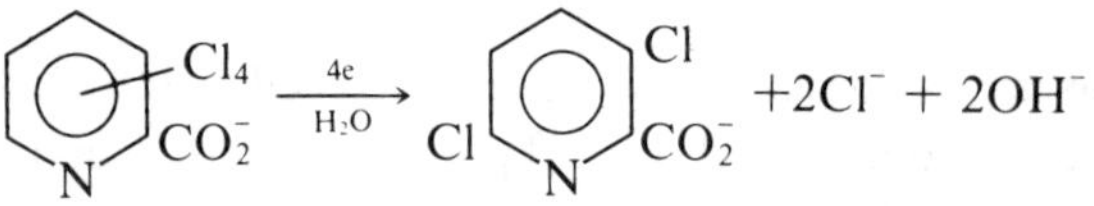

$$4OH^- \longrightarrow 2H_2O + O_2\uparrow + 4e$$

Although the above reactions are known to be as written, we assume here that they are only plausible expectations. In this electrolytic reduction we wish to achieve a very selective reduction: namely, electrohydrogenolysis at the 4 and 5 carbon-chlorine bonds of the starting material. We have two reasonable working hypotheses in favor of this selective reduction:

1   The tetrachloropicolinate ion has a permanent dipole moment that might favor an orientation of the ion at the electrode surface such as to expose the 4 and 5 positions of the ring to the cathode:

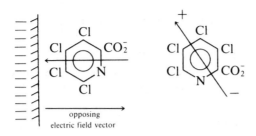

2 This organic ion would probably have least electron densities at the 4 and 5 carbons of the ring, which again would favor electron addition at these positions, regardless of any orientation effects as assumed above.

With regard to the anodic reaction it is very safe to expect that, in the presence of a large excess of hydroxyl ion, the predominant anodic reaction would be evolution of oxygen. These are logical assumptions in our preliminary attempts for the undertaken objective. Before proceeding any further, however, it would be most advisable to do some polarographic and voltammetric work. A polarogram of **1** in 5% aqueous NaOH is shown in Fig. 4.3.

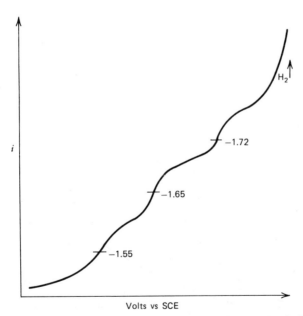

Figure 4.3 Polarogram of tetrachloropicolinic acid in 5% NaOH solution, mercury cathode.

Addition of some **2** to the solution in the polarographic cell augments the third wave. This is a good indication that the first two waves are probably due to the overall reduction as written. Attempts to carry out the electrolysis on a preparative 10 g scale, using a mercury cathode and a graphite anode, in both divided and undivided cells, have resulted in only 20 to 30% yield of the desired 3,6-dichloropicolinic acid. This rather low yield is most probably due to the closeness of the reduction potentials of the second and third wave. In this synthesis we expect the yield to be more than 90% if the working hypotheses are correct. Further voltammetric work has shown that among many electrodes (Fe, Cu, Ni, Pt, C, Zn, Sn, Ag, Pb) only at a *spongy* silver cathode was a reduction wave developed. Voltammograms with the spongy silver electrode are shown in Fig. 4.4. The spongy silver cathode was prepared by anodization, as described below, for both voltammetric and preparative scale electrolysis.

4.4.1   The Procedure

(The procedure given here was the result of many laboratory experiments that established the optimum conditions for this electrolysis.)

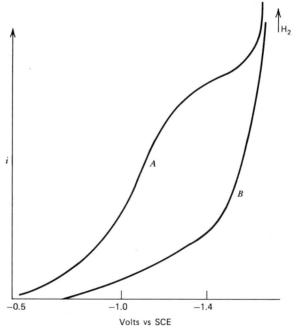

Figure 4.4   Curve *A*: voltamogram of tetrachloropicolinic acid in 5% NaOH solution, spongy silver cathode. Curve *B*: Blank.

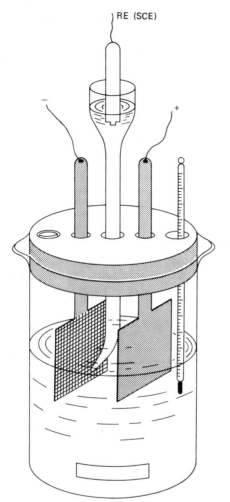

Figure 4.5   Electrolysis cell: spongy silver (screen) cathode, graphite plate anode.

A three-electrode cell is used without diaphragm. Electrolysis is carried out at constant cathode potential, near −1.3 V versus the SCE, at 25 to 35°C. A schematic representation of the cell is shown in Fig. 4.5. The cathode is a 20-mesh silver screen, 5 × 5 cm, and the anode a plate of graphite of equal dimensions. (The anode could also be silver oxide on silver.) The solution to be electrolyzed is composed of 150 ml distilled water, 12 g of reagent grade NaOH, and 10 g of tetrachloropicolinic acid (the last to be added in several steps, since the solution tends to form very stubborn foam if the acid is added all at once). The cathode and anode assembly is held together with neoprene rubber and is positioned in the center of a 400 ml glass beaker. Stirring of the solution is effected by a magnet bar at maximum speed. The cell is placed in a water bath thermostated at 25 to 35°C. Before adding the organic acid, the silver electrode is made the anode, starting from 0 V and gradually increasing the potential up to +0.6 V. This anodization procedure takes about 3 min, most of the time at the higher potential. As the anodic current begins to fall due to the formation of silver oxides at the surface of the electrode, the potential is set at −1.3 V and is kept there until the current reaches a steady value (ca 0.3 A). A current–potential curve is then obtained that, qualitatively, reproduces the voltammetric blank curve. The potential of the silver cathode is adjusted to −1.3 V and is kept there automatically by the potentiostat (Princeton Applied Research, Model 371). About 2 g of tetrachloropicolinic acid, in powder form, is macerated with a small portion of the solution from the cell and is added to the cell. The current rises from 0.3 to 3 A as a result of the reduction of the organic acid (sodium salt). The entire amount of the organic acid, 10 g, is thus added in 2 g portions over a period of 8 hr. When all the acid is converted to the dichloro compound, the current has fallen almost to the initial value of ~0.3 A. The electrolysis is completed, and the product is recovered by acidifying the electrolyzed solution with 25 ml conc. HCl, whereupon copious crystallization of the product occurs. Filtration and drying of the product thus isolated has given 7 g of 98% purity of the desired compound 3,6-dichloropicolinic (~92% yield). Identification of the recovered product is made by comparison with an authentic sample using spectroscopic (IR, NMR) and chromatographic procedures. The impurities found have been trichloropicolinic acid and a trace of monochloropicolinic acid.

It was found that periodic activation in situ of the cathode by anodization (every ~3 hr), as described above, is very beneficial. Also, best reproducibility is obtained by cleaning the used silver cathode after every run with 1 : 1 conc. HCl/H₂O. This procedure has been found to remove traces of electrodeposited metals, Fe, Cu, Ni, Pb, which are present in the materials used, and which apparently poison the cathode and make it less selective. When such impurities have been added to the solution in the cell, either overreduction was very extensive, defeating the intended purpose of the electrolysis, or the reduction was completely hindered. The anodization procedures (as determined by X-ray fluorescence analysis) removes most metallic impurities except copper, which apparently has the worst effect on the silver cathode.

It would now be reasonable to postulate that two forces are largely responsible for this very selective electrohydrogenolysis: (1) *specific adsorption* forces on the spongy silver; and (2) orientational electrostatic forces favoring electron transfer to 4 and 5 carbons of the ring.

It should be emphasized that no reduction is possible on a shiny silver surface. This fact very strongly suggests the operation of specific adsorption forces on the spongy silver surface. Moreover, the observed apparent half-wave potential with the spongy silver is at -1.1 V while that with the mercury cathode is at -1.55 and -1.65 V (second wave), in qualitative accord with the expected effect of adsorption, in general, as regards the activation energy for the electron transfer steps. The adsorption decreases the activation energy because it influences the heat of activation of the electrochemical reduction (provided adsorption is the slowest step in the overall scheme of events; if desorption were the slowest step, reduction would be more difficult and a more negative potential, with its side consequences, would then be required).

D. Kyriacou, unpublished work
D. Kyriacou, F. Edamura, and J. Love, U.S. Patent 4,217,185 (1980).

EPILOG

As in all fields of applied knowledge, there is something ineffable in electroorganic synthesis that only practice in the laboratory can teach. The best way to become a proficient electrosynthetic chemist is to work initially with an expert electroorganic chemist. This may not always be possible. One, therefore, must resort to patient book reading and laboratory practice, starting with simple, already known, preparative examples from the available literature. The present availability of commercial potentiostats, polarographs, and introductory textbooks on electroorganic synthesis should enable an organic chemist to learn to apply the electrolytic method of synthesis quite well within a relatively short time (remembering, of course, that it took some years to become an expert in conventional organic synthesis!).

As far as laboratory-scale preparations are concerned, the electrolytic method of synthesis *could always* be considered as a convenient alternative. In many cases it can be superior to conventional laboratory methods, and in some cases it may prove to be the only method. When, however, the ultimate goal is to develop a plant process, the electrolytic method should be considered on the basis of its technical and economic advantages relative to chemical or catalytic methods.

The electrochemical process of synthesis should, in general, be recommended under one or more of the following situations:

1 When no chemical or catalytic method is known.
2 When the above methods exist but are economically or technically unfavorable, that is; (a) the method involves many steps; (b) it gives low yields; (c) it is a dissolving metal type reduction; (d) it requires exotic oxidizing or

119

reducing agents; and (e) recovery of unreacted materials and product is difficult.

3   When the product at the auxiliary electrode is also of commercial value as, for example, is the case with the propylene oxide process.

4   When reducing or oxidizing agents exist but are required in large amounts. In such cases, the possibility of electrogenerating the reagent in an electrolytic cell should be considered (homogeneous electrocatalysis).

5   When there are environmental pollution problems that might be minimized by the electrolytic method.

6   When reductive or oxidative dimerizations and organometallic compounds are desired (high concentration of reactants).

If an electrosynthesis has shown promise on a laboratory scale and large-scale development of the method is contemplated, it would always be beneficial to invite the electrochemical and chemical engineers to the laboratory and to decide on a coordinated effort for scale-up. Seldom, if ever, can the laboratory cell be offered as a model for plant production cells. The input of engineers and electricians would definitely be desirable and necessary at this stage of research.

The following general questions should be considered:

1   Type of electrolysis (ie, constant current or constant potential).

2   Type of cell, divided or undivided, and kind of diaphragm, if needed.

3   Kind of solvent and supporting electrolyte (flammability, toxicity, etc).

4   Cathode and anode materials and their durability.

5   Chemical and electrical yields.

6   KW-hr/unit weight of product.

7   Rate of the reactions, current densities, temperature control, and any special electrode treatment (activation and reactivation techniques).

8   Method of recovery of products, solvent and electrolyte.

APPENDIX

SOME FUNDAMENTAL ELECTROCHEMICAL CONCEPTS AND PRINCIPLES

A.1 The Principle of Electrolysis

If two electronic conductors (electrodes) are immersed in an electrically conducting liquid medium and a direct current is forced through the medium, chemical transformations take place according to Faraday's laws of electrolysis:

1 The amount of chemical change effected at electrodes in an electrolytic cell is directly proportional to the amount of electricity passed.
2 96,500 C (26.806 A-hr) cause a chemical change of 1 g-equiv. of a substance. The current in the solution part of the circuit is the movement of ions and is called *ionic* current. Positive ions, or *cations,* move towards the negative electrode, or the cathode, while negative ions, or *anions,* move towards the positive electrode, or the anode. In the wires and the electrodes the current is the movement of electrons and is called electronic current. Ohm's law applies to both types of currents.

Although the presence of ions is *necessary* for carrying the current through the solution, the ions do not necessarily have to be discharged at the electrodes, *provided some other substance is available at the electrode surface to exchange electrons with the electrode.* It must be understood that it is the chemical transformations at the electrodes, which are induced by the applied potential, that allow the passage of current. The current is the result of the applied electric field that induces the chemical transformations. If no chemical reactions were possible the electrodes would only become *polarized* by the applied voltage. A polarized electrode results when no electrons flow across the electrode–solution interface, that is, when no chemical transformations are occurring under the prevailing potential difference at the interface.

A.2   The Electrode Potential and the Electrified Interface (Electrical Double Layer)

The electrode potential and the electrified interface between an electrode and a solution is — to use a phrase of Bockris — *"the innermost essence of electrochemistry."* The electrode potential governs the course of the electrode reaction and is, therefore, of paramount practical importance. At every electrode-solution interface there arises spontaneously an electrical potential difference. This potential difference is a consequence of the formation of an *electrical double layer* of opposite charges. The physical situation resembles a charged electrical capacitor, but it is actually a very complex and not well understood phenomenon. For our purposes the electrode potential is a potential difference at the interface, between an electronic conductor, such as a metal, and a conducting liquid medium. This potential difference can be modified by an externally imposed potential so that a desired electrodic reaction may take place. Although this potential difference exists in an absolute sense, it cannot be measured experimentally. Only relative potentials can be determined experimentally. All electrode potentials are values expressed in volts relative to a standard reference electrode. The normal hydrogen electrode (NHE), which is assigned the value of 0 V at all temperatures, has been established as the point of reference for all standard electrode potentials. In practice various reference electrodes are used, namely, the saturated calomel electrode (SCE), the silver-silver ion or silver-silver chloride electrode, and others, depending on their stability in the electrolysis media and the purpose of the work. In electroorganic synthesis the three electrodes mentioned above are very frequently used. We cannot discuss the subject of reference electrodes in any detail here (see the general Bibliography).

From a thermodynamic aspect the potential of an electrode, E, expresses an *intensity* or an *electron pressure* factor in the equation:

$$\Delta G = -nFE$$

where ΔG is the free energy of the thermodynamically reversible electrodic reaction:

$$A + ne \rightleftharpoons B$$

and nF is a *capacity* factor expressed in coulombs.

Under zero net current and strict thermodynamic equilibrium the electrode potential could be given by the familiar Nernst equation:

$$E = E^0 - \frac{RT}{nF} \ln \frac{[B]}{[A]}$$

Unfortunately, as far as electroorganic synthesis is concerned, this equation is of no practical value and it may even mislead the inexperienced. In practice, where a substantial current must be drawn, it is the overpotential that is of value — kinetics rather than the assumption of reversible reactions as implied in the equation. In the vast majority of electroorganic reactions the experimentally measured electrode potentials and cell voltages differ greatly from the thermodynamic potentials calculated from the relations:

$$\Delta G = G_p - G_r$$
$$\Delta G = -nFE$$

where ΔG, G_p, and G_r refer to free energies of reaction, and of formation of products and reactants, respectively. Measured electrode potentials always include three more potentials or *overpotentials,* as shown by the equation:

$$E_a = E_r + \eta + E_c + IR$$

where E_a = actual potential measured against any reference electrode

E_r = reversible (equilibrium) potential

η = activation overpotential

E_c = concentration overpotential

IR = ohmic overpotential

There may also be a *liquid junction* potential (see standard physical chemistry books). For practical electrosynthetic work this type of potential is either neglected or minimized by using salt bridges. Usually, salt bridges are made in the form of narrow glass tubes filled with solutions of suitable salts, such as KCl, KNO$_3$, or tetraalkylammonium perchlorate, the last being most suitable for nonaqueous electrolysis media. One end of the bridge is in contact with the solution being electrolyzed and the other with the solution of the reference electrode. The junction potentials are thus canceled because they are numerically approximately equal but opposite in sign.

The *activation overpotential* is defined as

$$E_a - E_r = \eta$$

This overpotential is inherent in the electron transfer act. It represents the excess potential necessary for the electron to surpass the energy barrier for a particular electrodic reaction. We comment more on this later. The *concentration* or *polarization* overpotential, E_c, is caused by the depletion of the electroactive species at the electrode surface as a result of the electrodic reaction. The *ohmic overpotential* depends on the current density and the resistance of the solution between the surface of the working electrode and the tip of the Luggin capillary or the tip of the reference electrode. (Ideally, no current should flow through the reference electrode; therefore, a high impedance potential measuring device is used to measure electrode potentials. Commercially available potentiostats are now built so that the organic experimenter need not worry about such purely electrical refinements.)

The ohmic overpotential can be minimized by placing the tip of the reference (Luggin capillary) electrode as close as possible to the surface of the electrode under study. The concentration overpotential can be minimized by good agitation of the solution so that the thickness of the diffusion zone, at the electrode–solution interphase, can be as small as possible. Under such conditions the major contribution to the overpotential for an electroorganic reaction would be from the activation overpotential. Only when this is true will current–potential curves give a meaningful kinetic picture of the electrochemical reaction (Tafel line). It should be clear that, for all practical electrochemical processes, it is always necessary to apply an overpotential, otherwise the reaction will not be possible. In general, the more reversible the electrodic reaction, the lower the overpotential needed for the reaction to be driven to completion. Obviously, if the reaction is reversible, a potential above the observed half-wave potential, $E_{1/2}$, should be applied to drive the reaction towards completion, whereas for an irreversible reaction any, in principle, potential above the minimum (decomposition) potential is sufficient to drive the reaction all the way, although at an impractically slow rate. Perhaps this question could be raised: since electroorganic reactions do not obey the Nernst relation, why do we need to know the irreversible electrode potential? Why not simply the cell voltage? In practice, we only want to know and apply the proper cell voltage for a particular cell reaction. However, the proper cell voltage is the sum of the anodic and cathodic *actual* potentials plus the *IR* drop of the *entire* system. Thus when we say that a given organic substance is reduced at a potential of −1.0 V, for example, we mean that we have applied the proper cell voltage, as a result of which the cathode potential has attained the desired value of −1.0 V relative to the chosen reference electrode. We have actually applied an overpotential that lowers the activation barrier for the reaction and makes its rate practical. The total cell voltage is independent of any reference electrode and may vary, depending on cell geometry, the *IR* drop, and the anodic and cathodic reaction potentials E_a and E_c. It is the energy barrier at the single electrode–solution interface that the electron must overcome, and, phenomenologically, it is the potential difference at this interface that affects this energy barrier. This single, relative, electrode potential must be known in order to best reproduce the results and to guide the reaction on its proper course.

The practical importance of the electrode potential and the structure of the double layer can be appreciated from the following considerations:

1   The electron exchange reaction occurs at the electrified electrode–solution interface and its rate is governed by the potential difference therein.

2   The physical and chemical properties of this very narrow region (ca 10 Å) are different from those prevailing in the bulk solution. The electrical fields in this region may be as strong as 10^7 V/cm. In such enormous fields no organic molecule can remain *indifferent.*

3   The building blocks of the electrical double layer on the solution side are usually the ions of the supporting electrolyte (usually, because there are cases where ions other than those of the supporting electrolyte may drastically influence the structure of the double layer). Therefore, both the concentration and kind of ions of the electrolyte would variously affect the rate of the electrochemical reactions and the nature of the final products.

It is useful to discuss further, at this point, the concept of electrode potential and overpotential. Physically, if we could measure the electrode potential as a potential difference above the potential that would have existed at perfect thermodynamic equilibrium — zero net current, no net chemical reaction — it could then be stated that the electrode potential represents the electronic energy that must be given to the electrons in the electrode phase in order to excite them to new energy levels above those existing at equilibrium. These new *electronic levels* would be able to overlap with electronic levels in the solution phase, and a net electron transfer would be possible from the electrode to the solution for cathodic reductions, and from solution to the electrode for anodic oxidation. Such events must happen on both the anode and cathode of an electrolysis cell. Cathodic and anodic chemical reactions would then occur, giving the overall cell reaction. In pure electrochemistry we speak of *Fermi electrons* and *tunneling* of electrons at the electrified interface. The physical meaning, therefore, of the electrode potential, or better the *overpotential* — since we are considering a potential above the thermodynamically reversible potential — is an expression of the excess energy in units of electron volts given to the electrons of a particular electrode dipped in a particular electrolysis environment. This energy disturbs the thermodynamic equilibrium and causes a net electron flow in one or the other direction at the electrode–solution interface. As a result of this electron transfer, chemical transformations are possible.

In practice, electrode potentials are measured with reference to some standard electrode. Electrode potentials thus expressed are not overpotentials, as defined above, although in practical work they are useful and sufficient. Relative electrode potentials of various electrodes can very easily be made equal by merely turning a knob in the potentiostat. However, their overpotentials may be, and usually are, different, for a given reaction, and they may not be equalized by simply readjusting the cell voltage. It is because of these differences in overpotentials that one electrode *works* for a particular reaction while another does not, although their potentials, as measured against the reference electrode, can be made to be exactly equal. Consider the hypothetical case in Fig. A.1. The difference ΔE in the relative potentials of two electrodes (curves A and B), both being tested for the same electroreduction,

$$Ox + e \longrightarrow Rd$$

is a difference of overpotentials, that is, of electronic energies that must be supplied to the electrons in the electrodes, in order to cause the reactions to proceed at the same rate (equal currents). This is the meaning and importance of the concept of overpotential. From current–potential curves such as those of Fig. A.1 we can select the most efficient electrode for an electrosynthesis.

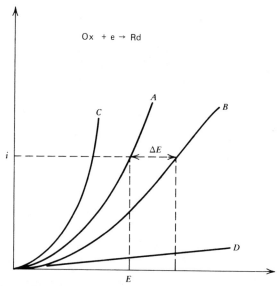

Figure A.1 Illustration of electrode overpotentials for a given electrodic reaction. Curves *A, B, C,* and *D* corresponding to different electrodes. *E* is relative electrode potential against the same reference electrode.

We can say that the electroreduction at the first electrode (curve *A*) is fairly rapid (easier) than at the second electrode (curve *B*), and very rapid at the third electrode (curve *C*). It is not feasible at the last electrode (curve *D*).

In most instances the experimental curves are only anodic or cathodic branches because most electroorganic reactions are irreversible thermodynamically (remember the distinction between chemical and thermodynamic reversibility).

Reversible and irreversible electrochemical reactions are to be viewed in relation to the experimental time scale, and in regards to whether we have in mind the primary electron transfer step or the final electrochemical reaction products:

$$X \pm e \rightleftharpoons [X] \longrightarrow \text{products}$$

Primary electron transfers appear frequently to be reversible steps depending on the time scale of the observation (as in cyclic voltammetry). It is the following chemical steps that usually render the reaction irreversible.

A.3   The Cell Voltage

The cell voltage is the measured potential difference between the cathode and anode. This potential difference, which can be measured experimentally by any ordinary voltmeter, is independent of the reference electrodes used for the determination of the single electrode potentials, provided the reference electrode is the same for both anode and cathode. The total, terminal potential difference, or voltage, across any cell includes three potential terms:

$$E_{total} = E_a - E_c + IR$$

where E_a and E_c represent the *measured* anode and cathode potentials against a reference electrode, and IR represents the *entire* ohmic potential drop of the system (solution, wires, diaphragms, liquid junctions, films).

Although in the laboratory the total cell voltage may be of little concern, it is this voltage that enters into the economics and technology of industrial production, since the cell voltage will determine the energy (watt-hours) per unit weight of products and the cost and kind of the electric power equipment to be used in the plant.

In order for any electrolytic cell reaction to be possible, a certain voltage must be applied to the cell. The minimum voltage needed for the cell reaction to proceed at a reasonable rate is called the *decomposition voltage.* A better term, as far as electrosynthesis is concerned, would be *reaction voltage.* As the applied voltage is raised above a certain minimum value, the current through the cell begins to rise gradually and then more steeply, as shown in Fig. A.2.

When current flows through a cell at least two reactions must occur, one at the anode and the other at the cathode:

$$A \longrightarrow B + ne \qquad \text{anodic}$$
$$C + ne \longrightarrow D \qquad \text{cathodic}$$
$$A + C \longrightarrow B + D \qquad \text{cell reaction}$$

The cell voltage for an electrosynthesis can easily be found by plotting current versus cell voltage, as illustrated in Fig. A.2. It is important not to confuse cell voltages with cathode or anode potentials as determined by polarography or voltammetry, or with the potentials of the single electrodes in a cell as measured against the reference electrode. The cathode or anode potentials are usu-

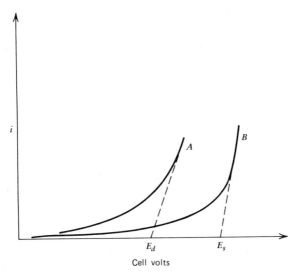

Figure A.2   Current–cell voltage curves. Curve A: Curve with organic substrate. Curve B: Blank or background curve. E_d: Decomposition voltage (reaction of organic substrate). E_s: Solvent or electrolyte decomposition voltage.

ally 0 ± 3 V, whereas the cell voltages are very often 2 to 20 V, and as high as 100 V with some poorly conducting media.

In Fig. A.2 it is assumed that the desired electrodic reaction is direct, and that it occurs at potentials within the electrochemical domain of the electrolysis medium, that is, before the potential at which the solvent or the electrolyte suffers electroreduction or electrooxidation. If an electrolytic reduction is indirect, such as an electrohydrogenation reaction or an oxidation catalyzed by electrolytically generated halogens, the current–voltage curves may correspond to the reactions:

$$2H_2O + 2e \longrightarrow H_2 + 2OH^-$$
$$2Br \longrightarrow Br_2 + 2e$$

The feasibility of these electrosyntheses would be difficult to infer from current–potential curves alone. An experienced polarographer may, sometimes, detect from carefully recorded polarograms a depolarizing effect by the organic substrate in such indirect reactions. The polarograms become less cathodic for reductions and less anodic for oxidations (see Section 1.7).

A.4   The Current–Potential Relationship

The practical equation for the rate of electrochemical reactions is Tafel's (1905) empirical equation:

$$\eta = a + b \log i$$

$$i = nF \frac{dc}{dt}, \qquad \frac{dc}{dt} = \text{reaction rate}$$

This equation has now been deduced from theoretical concepts. It relates the applied potential or overpotential, η, with the current, i, or the rate of the electrodic reaction. The constants a and b pertain to a particular reaction. The equation applies only when mass transfer processes and chemical reactions are faster than the electron transfer step, and when the current is dependent only on the activation overpotential η. This means that the electron exchange step must be the slowest of all steps in the electrode process.

The Tafel equation is a special form of the more general kinetic equation:

$$i = i_0 \left\{ \exp\left(\frac{-\alpha n F \eta}{RT} \right) - \exp\left[\frac{(1 - \alpha)}{RT} nF\eta \right] \right\}$$

where i = net current

i_0 = exchange current (see below)

η = overpotential, defined as $E - E^0$

E^0 = standard equilibrium potential when $C_O = C_R$
       for the electrode reaction:

$$O + ne \rightleftharpoons R$$

i_0 = $nFk_s A C_{(O \text{ or } R)}$

n = number of electrons

α = transfer coefficient

k_s = heterogeneous rate constant in cm/sec

A = electrode area

F, R, and T have their usual meanings.

(The general equation above was first proposed by Erdey-Gruz and Volmer (1930); see ref. 1.)

The electrochemical activation energy is defined as

$$\Delta G_{el}^{\ddagger} = \Delta G_{ch}^{\ddagger} - \alpha nF(E^0 + \eta)$$

The electrochemical activation energy is the difference:

$$\Delta G_{ch}^{\ddagger} - \alpha nF\eta, \text{ if } {}^{!}E^0 \equiv 0,$$

from which the general current–potential relation, as expressed above, can be obtained. In theory, therefore, barrierless kinetics can be possible in electrochemical reactions (when η assumes a value so that the $\Delta G_{ch}^{\ddagger} - \alpha nF\eta = 0$). In reality, however, this situation can never be achieved because other physical barriers arise that impose their own limits on the reaction rates.

These equations show a very important fact: exponential dependence of current on the activation overpotential. The activation overpotential is a measure of the change in the chemical free energy of activation for the electron transfer step:

$$\eta = \Delta\Delta G^{\ddagger}$$

Thus a change in the electrode potential of 1 V corresponds to a change in $\Delta G^{\ddagger}$ of (α) (23 kcal/g-equiv.). The transfer coefficient, α, is a number, usually between 0.3 and 0.7. It is often assumed to be 0.5.

When the overpotential η is measured as the difference of the measured electrode potential and the *standard* equilibrium potential, E^0, when by definition $C_O = C_R$, the value of k_s is given as

$$k_s = \frac{i_0}{nFAC_O} = \frac{i_0}{nFAC_R}$$

Tafel lines obtained by graphing η versus log i are useful in the study of electrode reaction mechanisms. The Tafel slope, b, can be deduced from the theoretical kinetic equation given above, and is thus given by this equation:

$$b = \frac{2.303RT}{\alpha nF}$$

so that the value of b can be a valuable diagnostic criterion for the mechanism of an electrodic reaction. The exchange current can also be obtained from the Tafel line (intercept on the current axis, when $\eta \equiv E - E^0 = 0$).

The kinetic order of an electrodic reaction is defined as

$$\text{order} = \left(\frac{d \log i}{d \log C}\right)_E$$

where the current, at a constant potential E, represents the rate of the reaction and C is concentration of the reactant at the electrode surface.

A.5 The Meaning of the Exchange Current

In every electrochemical reaction the current that is experimentally observed is a *net* current, that is, a difference of two opposing currents at each electrode:

$$|i_c| - |i_a| > 0 \qquad |i_c| - |i_a| < 0$$

The electrode potential at which the opposing currents are numerically equal is the potential of electrochemical equilibrium. This means that the electroactive species at the electrode surface accepts and releases electrons at the same rate so that there is no *net* current (no current in the external wires). The equilibrium is a dynamic one, as are all chemical equilibria. As many electrons cross the interface in one direction as in the other, so that

$$i_c = i_a = i_0$$

(It must be understood that zero current does not necessarily mean an equilibrium condition. In most cases the system is too slow, and appears *at rest*.) The exchange current, i_0, cannot be observed experimentally but, as was indicated previously, it can be obtained from Tafel lines at $\eta = 0$. Exchange currents vary from about 10^{-12} to 10^{-2} A/cm^2. Reversible reactions are characterized by high exchange currents, whereas irreversible reactions are characterized by low exchange currents, the higher the exchange current the more reversible (mobile) the reaction. The exchange current is the result of an inherent tendency of an electrode–solution interface to exchange electrons. The externally imposed potential on the interface modifies this tendency of the system so that more electrons cross the interface in one direction than in the other, and hence a *net* reaction occurs and a *net* current appears in the external circuit.

A.6 The Potential of Zero Charge

There is a relative potential at which an electrode has zero charge on its surface. This potential is known as the *potential of zero charge* (PZC). It means that the electrode surface has no excess of lack of electrons. At this neutral surface, organic neutral molecules tend to be adsorbed at the electrode surface more strongly than they do at other potentials. Ionic species tend to do the opposite. Positive species are adsorbed most on the negative side of the PZC, while negative are adsorbed on the positive side, as would normally be expected from the operation of electrostatic forces. The PZC depends on both the electrode material and the composition of the surrounding liquid medium. Because many, if not most, electroorganic reactions proceed through adsorbed species, the PZC is of practical and theoretical significance. Although its determination is difficult and its effects cannot be predicted, the PZC should be, conceptually at least, considered in electroorganic synthesis research.

A.7 Adsorption at Electrodes

In electroorganic synthesis the phenomenon of adsorption (or electrosorption, when it is influenced by the electrode potential) is of very great practical importance. Indeed, many electrodic reactions cannot occur without adsorption of reactants at the electrode surface. The electrocatalytic properties of an electrode are related to its surface morphology. Adsorption and catalysis are interrelated and further influenced by the electric field or the electrode potential.

It can be shown that the current, that is, the rate of an electrochemical reaction, is influenced by the free energy of adsorption of the reactant onto the electrode surface:

$$i = \text{constant} \cdot \exp \frac{-(\Delta G_{ch}^{0\ddagger} + \Delta G_{ad}^{0\ddagger} - \alpha F \eta)}{RT}$$

The current is an exponential function of the sum of three terms, which constitute the electrochemical activation energy for the reaction:

$$A \pm e \rightleftharpoons B$$

where ΔG_{ch}, ΔG_{ad}, and $\alpha F \eta$ represent the chemical, the adsorption, and the electrical (overpotential) parts of the total activation energy, respectively. The tremendous role of the overpotential η becomes clear. It is entirely possible to change the applied cell voltage, and hence the value of η, at will. The rate of the reaction can therefore be made an effect of zero activation energy, when the overpotential assumes a value so that

$$\eta = \frac{\Delta G_{ch}^{0\ddagger} + \Delta G_{ad}^{0\ddagger}}{\alpha F}$$

The importance of the ΔG_{ad} also, and hence of the adsorptive nature of the electrode, is obvious from this theoretical equation.

We have only touched upon the effect of adsorption on the rate of electrochemical reactions. It can be perhaps inferred from the foregoing discussion that the effect of adsorption is as if the electron transfer reaction tends to occur at an *underpotential* rather than an overpotential, owing to the gained energy of the species in the adsorbed state relative to the energy of the unadsorbed species. Underpotential deposition of metallic ions is a recognized phenomenon. However, with metals the advantage of underpotential disappears as soon as the first monolayer of deposited ions is formed. With organic molecules desorption occurs and the electroreduction or electrooxidation may thus continue under underpotential conditions (on this subject refer to ref. 2).

A.8   Electrocatalysis

By definition a catalyst is a substance that alters the rate of a chemical reaction without itself being consumed in the process. On this basis, all inert electrodes are catalysts, and all electrochemical reactions might be viewed as heterogeneous catalytic processes, under the influence of imposed electric fields. Because the electrode is a catalyst for the electron transfers, it is called an *electrocatalyst* and the phenomenon an *electrocatalysis.*

Electrodes have different catalytic activities, depending on the electrode material, the surface morphology, and the treatment, in general, of the electrode.

Organic substances tend to be adsorbed on electrode surfaces and are susceptible to heterogeneous electrocatalysis (direct electron transfer) at the adsorbed state. By virtue of this type of catalysis special reactions and products may be possible that would be difficult or impossible by other synthetic methods.

With regard to the rate of the electrodic reaction as influenced by adsorption on an electrocatalytic surface, two general deductions can be made quite intuitively:

1   The stronger the tendency for adsorption, the easier the electron transfer (ie, less cathodic potentials for reductions and less anodic potentials for oxidations), *provided* adsorption is the rate determining step in the overall scheme of events.

2   If desorption of the products is the slowest step, then the opposite should be expected to occur.

Polarographic and voltammetric current–potential curves usually can indicate such adsorption effects and can thus guide the worker in the search for electrode materials for a particular synthesis.

If changing the electrode material or the surface morphology of an electrode causes significant changes in the current at a fixed potential, it is very likely that adsorption forces are operating. Thus a substance may be electroactive at one electrode and not at another, even of the same material, if the electrode surfaces are microscopically different.

Electrodes may become *passive* in the course of an electrolysis or in subsequent runs. Some way must then be found to reactivate the electrode. Often reversing the polarity in a certain way restores the activity of the electrode. This can be done *in situ* or *ex situ*. Passivation of electrodes is due to various causes. Nonconductive films or layers of species retarding the electrodic reactions may be formed at the electrode. These films may be organic, metallic, solid, or viscous liquids. Often metallic nonconductive oxides passivate some anodes. Electrocatalytic surfaces may sometimes be metastable structures, requiring periodic regeneration.

There is, apparently, no one method for depassivating, or activating, an electrode. Polarity reversals, chemical and mechanical treatments, and ultrasonic vibrations are the methods most frequently used.

Some types of *filming* (as is probably the case with the so-called chemically modified electrodes) on the electrode surface may prove to be very desirable, provided the film allows the passage of current. Inert species, such as the ions of a supporting electrolyte or other electrically charged impurities, may preferentially become adsorbed or heavily concentrated at the electrode surface. The course of the reaction might be more or less affected in such cases; sometimes the effect is very drastic. It is known, for example, that acrylonitrile in aqueous media hydrodimerizes to adiponitrile when the electrolyte is a quaternary ammonium salt, whereas if the salt is an alkali metal salt the main product is propionitrile (Baizer, Monsanto process).

A.9   Electrogeneration of Catalysts

The discussion above has been about heterogeneous electrocatalysis. Electrogeneration of catalysts for homogeneous catalysis is another very promising field in electroorganic chemistry. The electrolytic reaction may involve only the generation and regeneration of the catalytic substance. Reversible redox systems, usually inorganic, have been used successfully. These electrochemical processes are indirect oxidations or reductions where the oxidizing or reducing agent is regenerated *in situ*. In symbols,

$$M^{3+} + e \longrightarrow M^{2+} \qquad \text{at cathode}$$
$$M^{2+} + R \longrightarrow R' + M^{3+} \qquad \text{in solution}$$

Metallic ions, and other inorganic and organic species in various oxidation states may thus be used as catalysts for certain reactions. We may visualize many possibilities on the basis of known chemical redox reactions. For example, some possible redox systems would be the following:

$$Fe^{2+}/Fe^{3+}$$
$$VO^{2+}/V^{3+}$$
$$Ce^{3+}/Ce^{4+}$$
$$Co^{2+}/Co^{3+}$$
$$Sn^{2+}/Sn^{4+}$$
$$Cr^{2+}/Cr^{3+}$$

Although the catalytic activity would be expected to be directly related to the redox potentials of the couples, in practice this may not be so because it is not thermodynamics but kinetics that determine the effectiveness of the redox couple.

References

1   J. O'M. Bockris and A. K. N. Reddy, *Modern Electrochemistry,* Vol. 2, Plenum, New York, 1970.
2   S. Trasatti, in H. Gerischer and C. Tobias, Eds., *Advances of Electrochemistry and Electrochemical Engineering*, Vol. 10, Wiley, New York, 1977, p. 255.

SOLVENTS AND SOLVENT MIXTURES COMMONLY USED IN ELECTROORGANIC SYNTHESES

The solvents used in electroorganic syntheses are generally classified as protic, or high proton availability, and aprotic, or low proton availability, solvents. In the first class belong water, alcohols, acids, and in the second class solvents such as acetonitrile, dimethylformamide, sulfolane, pyridine and several others. Aprotic, nonpolar solvents are, of course, many but they can be used only in mixtures with polar solvents, protic or aprotic. Solvents are also classified as nucleophilic or electrophilic (e.g., acetonitrile, acetic anhydride).

In many cases the electrolysis medium consists of only one solvent and the supporting electrolyte. There are cases where mixtures of solvents must be used to enhance solubility of the reactants, to form the needed chemical medium (solvent participation in the overall reaction), or to aid in the selectivity of the reaction. In this section some of the most commonly used solvents and solvent mixtures for synthetic work are briefly discussed. (For a description of solvents see ref. 1).

Water

Several oxidations and reductions are possible using water as a medium. Water is the best solvent (dielectric constant 80) as regards to ionizing ability, and hence conductivity, and, in general, *safety* and economics of the electrolytic process. Practically all types of salts, acids, and bases (and most types of reference electrodes) can be used in aqueous media. Organic solids that are insoluble in water cannot, of course, be electrolyzed. However, such solids can be dissolved in some organic solvent and the resulting mixture can then be used for the electrolysis as an emulsion. Organic liquid substances can be directly used in water and electrolyzed under vigorous agitation, thus often obviating the need for an additional solvent. The potential range of water depends on pH, supporting electrolyte, and electrode material. Negative potentials up to −2.5 V and positive potentials to +1.5 V versus the SCE can be reached using the proper electrolyte and electrodes.

Acetonitrile

This organic solvent is very frequently used in organic synthesis. It has very good solvent properties, dissolves most organic substances, and is miscible with water. Its disadvantages are that it is toxic and flammable.

Dielectric constant: ~37

Liquid range: -41 to $82°C$

Approximate potential range: $+2.4$ to -3.5 V

Electrolytes: lithium halides, quaternary ammonium salts (perchlorates, halides, tetrafluoroborates)

Reference electrodes: $Ag/Ag^+NO_3^-$, SCE, $Ag/AgCl$

The last two are used with salt bridges, since they are unstable in acetonitrile.

For many synthetic reactions acetonitrile can be used without purification. For polarographic and voltammetric studies it must be purified, since it contains several impurities that usually cause difficulties in the interpretation of current–potential curves.

N,N-Dimethylformamide (DMF)

This solvent is used to about the same extent as acetonitrile. It is toxic and flammable.

Dielectric constant: ~37

Liquid range: -61 to $153°C$

Approximate potential range: $+1.6$ to -3.0 V vs SCE

Electrolytes: as with acetonitrile

Reference electrodes: aqueous SCE, cadmium amalgam/sodium perchlorate, Hg/LiCl, Hg/n-BuNI.

For synthetic work it may be used without purification. For polarographic or voltammetric studies it needs purification.

Methanol

Methanol, alone or in mixtures with water and other solvents, is a good solvent for many electroorganic syntheses. It dissolves several salts and is a good solvent for alkali hydroxides. It is toxic and flammable.

Dielectric constant: ~33

Liquid range: -98 to $65°C$

Potential range: -2.0 to $+1.3$ V $(LiClO_4)$ (not well defined)

Electrolytes: alkali hydroxides, halides, perchlorates, quaternary ammonium salts

Reference electrodes: aqueous SCE, Ag/Ag$^+$

Ethanol

It is best used in mixtures with water. In general, it behaves like methanol.

Tetrahydrofuran (THF)

Because its dielectric constant is low, ~7.5, it is preferably used with water or some polar solvent for preparative work. It is toxic and flammable.

Liquid range: −65 to 66°C

Potential range: −3.5 to +1.8 V

Electrolytes: quaternary ammonium salts, lithium perchlorate

Reference electrodes: aqueous SCE, Ag/Ag$^+$

Dioxane

Because its dielectric constant is very low, 2.2, dioxane is inadequate by itself, but it can be used when mixed with water or other polar solvents. It is toxic and flammable.

Dimethyl Sulfoxide

This solvent is readily absorbed through the skin but in itself is not toxic. However, some strange physiological effects have been reported for this material, and although it may not be toxic its rapid absorption through the skin may introduce other toxic materials into the body.

Dielectric constant: 46

Liquid range: 18 to 190°C

Potential range: −3.0 to +0.7 V

Electrolytes: alkali salts, quaternary ammonium salts

Reference electrodes: aqueous SCE, with salt bridges

Sulfolane

This is a very convenient nontoxic, nonflammable, water-soluble organic solvent. Almost all kinds of electrolytes can be used with sulfolane.

Dielectric constant: 44

Liquid range: 27 to 285°C

Potential range: −2.0 to +3.0 V

In many preparative electrolyses aqueous-organic mixtures, homogeneous or heterogeneous, are employed. Organic solvents most often used in such mixtures are the lower alcohols, carboxylic acids, various amines, ethers, pyridine, sulfolane, propylene carbonate, and several other polar and nonpolar liquid substances. Homogeneous media may turn heterogeneous depending on the type and amount of supporting electrolyte. Sometimes this is desirable, especially when the organic reactant is not sufficiently soluble in the homogeneous medium but is more soluble in the phased-out organic layer. Heterogeneous media would require vigorous stirring, and when constant potential electrolysis is performed the position of the reference electrode tip (Luggin) would need special care and consistency if meaningful monitoring of the electrode potential is to be achieved.

For electrochemical studies and very special syntheses, solvents such as methylene chloride, hexamethylphosphoramide, dimethoxyethane, amines and amides, liquid ammonia, liquid sulfur dioxide, and several other solvents can be used. Some solvents such as ammonia, amines, and hexamethylphosphoramide are good media for *solvated* electrons. Homogeneous reduction by solvated electrons can be possible in such media.

In some cases sulfuric acid of various concentrations, fluorosulfonic and trifluoroacetic acids are very valuable solvents.

References

1   C. K. Mann, in A. J. Bard, Ed., *Electroanalytical Chemistry*, Vol. 3, Marcel Dekker, New York, 1969.

APPROXIMATE POTENTIAL RANGES FOR REDUCTION AND OXIDATION OF ORGANIC COMPOUNDS

These ranges are inferred from polarographic and voltammetric data in various media. They are referred to SCE and are only intended to give a general idea of potentials to be expected in electrosynthetic work.

Reduction	Range (V)
Activated olefins	-1.5 to -3.5
Aldehydes	-1.5 to -2.0
Aromatics	-1.5 to -3.0
Azo	$+0.05$ to -0.1

Reductions	Range (V)
Azoxy	−0.5 to −0.1
Diazo	−0.3 to −1.0
Esters (activated)	−1.0 to −2.0
Halides	
Bromides	−0.5 to −2.0
Chlorides	−0.5 to −3.0
Iodides	−0.3 to −1.5
Nitriles (aliphatic)	electrohydrogenate at low hydrogen overvoltage electrodes
Nitriles (aromatic)	−1.5 to −3.5
Nitros	−0.5 to −1.0
Nitrosos	−0.5 to −2.0
Sulfones	−1.5 to −2.5
Hydrazines	−1.5 to −2.0
Hydroxylamines	−0.5 to −1.5

Oxidations	Range (V)
Alcohols	+1.5 to +2.0
Amides	+1.5 to +2.5
Aromatic amines	+1.0 to +2.5
Aromatic hydrocarbons	+1.0 to +2.5
Azos	+1.5 to +2.0
Carboxylates	+2.0 to +2.5
Diazos	+1.0 to +2.0
Ethers	+1.0 to +2.0
Hydrazos	−1.0 to +0.5
Ketones	+1.0 to +2.0
Phenols	+0.5 to +1.0
Sulfides	+0.5 to +1.5

Note: A comprehensive list of oxidation and reduction potentials, $E_{1/2}$ is included in N. L. Weinberg, Ed., *Technique of Electroorganic Synthesis*, Part II, Wiley-Interscience, New York, 1975.

AVERAGE OVERVOLTAGE OF HYDROGEN ON VARIOUS CATHODE MATERIALS

Cathode	Overvoltage (V)	Cathode	Overvoltage (V)
Platinized platinum	0.03	Copper	0.67
Tungsten	0.27	Iron	0.71
Smooth platinum	0.29	Graphite	0.77
Antimony	0.43	Aluminum	0.80
Gold	0.48	Mercury	0.89
Nickel	0.56	Tin	0.92
Palladium	0.59	Zinc	0.94
Silver	0.62	Lead	1.00
Carbon	0.64	Cadmium	1.22

Reprinted with permission from F. D. Popp and H. P. Schultz, *Chem. Rev.*, **62**, 22 (1962). Copyright 1962, American Chemical Society.

PREPARATION OF SPONGY ELECTRODE SURFACES

Generally speaking, all metallic electrode surfaces become *spongy* upon prolonged use, especially if they are used as anodes. Spongy electrodes not only have high surfaces, hence allowing more current to pass per apparent unit area, they also possess, to various degrees, other desirable characteristics related to adsorption, electrocatalysis, and *actual* current densities. The platinized platinum electrode, for example, because of its microscopic surface morphology, has a greater electrocatalytic effect than a smooth platinum electrode for the hydrogen evolution reaction. Smooth platinum, therefore, requires higher overpotential than the platinized platinum for reduction of water to hydrogen and OH^- ion. Platinized platinum can be prepared by electroplating platinum from 1 to 3% chloroplatinic acid solutions.[1]

Lead dioxide anodes are spongy electrodes that can be readily prepared by anodizing a lead electrode in aqueous sulfuric acid solutions, or by electroplating lead dioxide on lead, iron, nickel, graphite, and other substrates.[2] Lead nitrate solutions are commonly used. Obviously, lead dioxide as such is not practical for use as cathode (if used it will be rapidly converted to metallic lead in the process). Silver and nickel oxide anodes can be readily prepared in strong sodium hydroxide solutions using the silver or nickel electrode as anodes. Reversing the polarity several times produces very spongy surfaces. Such surfaces are very effective for both cathodic and anodic electrochemical reactions. Spongy silver surfaces prepared in aqueous sodium hydroxide solutions are very effective cathode surfaces for reduction of various carbon-halogen bonds.[3]

It should be noted, in general, that the manner of anodizing and cathodizing (i.e., potential excursion, time, medium composition) may determine the resulting surface character of the electrode and hence its electrocatalytic activity for a particular reaction or type of reactions.

References

1   R. G. Bates, *Determination of pH*, Wiley, New York, 1964, p. 242; F. M. Feltman and M. Spiro, *Chem. Rev.*, **71**, 177 (1971).

2   J. P. Carr and N. A. Hamson, *Chem. Rev.*, **72**, 679 (1972).

3   D. Kyriacou, unpublished work; U.S. Patent 4, 242, 183 (1980).

GLOSSARY AND DEFINITIONS

Ampere (A)   1 A = 1 C/sec = 6.24 $\times 10^{18}$ electrons/sec = 1.036 $\times$ 10^{-5} g-equiv./sec.

Anode   The electrode that is connected to the positive pole of the dc power source and accepts electrons from the solution.

Anolyte   The solution surrounding the anode.

Auxiliary (counter) Electrode   The electrode at which the auxiliary reaction (usually not of direct synthetic concern) occurs.

Cathode   The electrode that is connected to the negative pole of a dc power source and donates electrons to the solution.

Catholyte   The solution surrounding the cathode.

Cell Voltage   The total measured voltage across the cathode and anode.

Coulomb (C)   1 C = 1 A-sec = 6.24 $\times 10^{18}$ electrons (defined as that amount of electricity depositing 0.0011180 g of silver from a solution of silver nitrate).

Electrode potential   The potential difference, in volts, as measured against a reference electrode, at an electrode–solution interface. It is a measure of the oxidizing or reducing power of the electrode in the particular electrolysis system.

Electrolysis   The technique by which chemical reactions take place in solution by the application of an electric potential difference (passage of current) between two electronic conductors (electrodes) in contact with the solution. When the current is kept constant the electrolysis is called *constant current electrolysis*, while if the electrode potential is kept *constant* the electrolysis is called *constant potential electrolysis*.

Electrolyte   The medium in which electrolysis takes place.

Equilibrium electrode potential   The electrode potential at which thermodynamic (Nernstian) equilibrium exists, that is, the potential at which the rates of electron transfer to and fro at the electrode–solution interface are equal and no net current or net chemical transformations are occurring. If the species involved in the electrode reaction are at unit activity, that potential is the standard potential of the reaction.

Faraday (F)   1 F = 96,500 C = 26.806 A-hr.

Half-Cell   In a divided cell the anolyte and the catholyte compartments constitute half cells. A half-cell reaction represents the single electrode reaction, i.e., $M \pm ne \rightleftharpoons M^{n\pm}$. Half-cell reactions are abstract concepts, because in reality only paired electrode reactions are possible, that is, two half-cells or two half-cell reactions constitute an electrolysis system or cell reaction. When a half-cell reaction in an electrolysis cell is thermodynamically reversible and the electrode potential remains constant during the passage of a small current (as with a relatively large mercury pool electrode in the presence of chloride ion), that half-cell may function as a reference electrode against which another half-cell reaction potential can be measured.

Joule (J)   1 J = 1 W-sec = 0.239 cal.

Ohm's Law $E = IR$ (E = potential difference, I = current, R = resistance). The ohm (Ω) is the unit of electrical resistance. A 62 ft length of 22-gauge copper wire has a resistance of about 1 Ω.

Overpotential (overvoltage)   The difference between the actual, measured, electrode potential and that at equilibrium (as would be if the Nernst relation were applicable). For the hydrogen evolution reaction the hydrogen overpotential refers to the cathodic reaction:

$$2H^+ + 2e \longrightarrow H_2$$

$$(2H_2O + 2e \longrightarrow H_2 + 2OH-)$$

while for oxygen evolution the oxygen overpotential refers to the anodic reaction:

$$2H_2O \longrightarrow 4H^+ + O_2 + 4e$$

$$(4OH^- \longrightarrow O_2 + 2H_2O + 4e)$$

Passivation of Electrode   Loss of the desired electrode activity either by alteration of the surface itself (i.e., deposition of metals or changes in the morphology of the surface) or by some film formation (as that of oxides and various nonconductive organic coatings).

Reference Electrode   A reversible half-cell (electrode) such as the normal hydrogen electrode (NHE), the saturated calomel electrode (SCE), and others, which are used in the measurement of relative electrode potentials.

Supporting Electrolyte   The substance providing the ions in the solution to carry the current through the solution part of the circuit.

Volt (V) $1 \text{ V} = 1 \text{ J/C} = 23.06 \text{ kcal/g-equiv.}$

Watt (W) $1 \text{ W} = 1 \text{ J/sec} = 0.239 \text{ cal/sec.}$

Faraday's Law and Cell Performance Concepts

$$m = \left(\frac{M}{nF}\right)(It), \qquad \text{constant current (100\% efficiency)}$$

$$m = \frac{M}{nF} \int_0^t I\,dt, \qquad \text{constant potential (100\% efficiency)}$$

where m = weight of substance reduced or oxidized, g
M = molecular weight of substance
n = number of electrons per molecule
$F = 96,500 \text{ C}$
I = current, A
t = time, sec

Current-Decay Relationship   This equation shows the relationship for constant potential electrolysis and 100% current efficiency:

$$i_t = i_0 e^{-kt}, \qquad k \approx \frac{DA}{\delta V}$$

where i_t = current at time t
i_0 = initial current
k = apparent rate constant
D = diffusion coefficient

A = area of electrode
V = volume of solution
δ = thickness of diffusion zone (~0.1 mm)

Total or terminal cell voltage, E_t

$$E_t = E_d + E_{conc.} + \eta + IR$$

where E_t = total voltage of cell as measured by a voltmeter connecting the
        anode and cathode
       E_d = the theoretical (Nernst) potential of the cell
    $E_{conc.}$ = the concentration overpotential
       η = the activation overpotential
       IR = total ohmic drop in the system

Current Density   Current per apparent (geometric) unit area of electrode.

Current Efficiency (CE) CE = theoretical A-hr/actual A-hr.

Voltage Efficiency (VE) $VE = E_d/E_t$.

Energy Efficiency (EE) $EE = (VE) \cdot (CE)$.

Energy per gram equivalent (EE/g-equiv.) EE/g-equiv. $= mnFE_t/M$.

A PARTIAL LIST OF MANUFACTURERS AND SUPPLIERS OF ELECTROCHEMICAL EQUIPMENT

Astra Scientific International, Inc.
P.O. Box 2004
Santa Clara, CA 95063

Brinkman Instruments
Cantiague Rd.
Westbury, NY 11590

Fairchild Instrumentation
974 E. Argues St.
Sunnyvale, CA 91361

Hewlett-Packard
Rt. 41, Starr Rd.
Avondale, PA 19311

Infotronics Corp.
7800 Western Dr.
Houston, TX 77055

Keithley Instruments
28775 Aurora Rd.
Cleveland, OH 44139

Kempco, Inc. (for rectifiers 10 V,
   100 A)
131-38 Sanford Ave.
Flushing, NY 11352

Perkin-Elmer
761 Main Ave., Station 10
Norwalk, CT 06856

Princeton Applied Research
P.O. Box 2565
Princeton, NJ 08540

Tectronix, Inc.
Box 500
Beaverton, OR 97075

Radio Corp. of America
Camden, NJ 08103

Note. No preference is implied. These sources happen to be known to the author.

GENERAL BIBLIOGRAPHY

R. N. Adams, *Electrochemistry at Solid Electrodes*, Marcel Dekker, New York, 1969.

M. Baizer, Ed., *Organic Electrochemistry*, Marcel Dekker, New York, 1973.

A. J. Bard, *Encyclopedia of Electrochemistry of the Elements* (Organic Section), Vols. XII ff, Marcel Dekker, New York, 1973.

J. O'M. Bockris and A. K. N. Reddy, *Modern Electrochemistry*, Vols. 1 and 2, Plenum, New York, 1970.

D. J. G. Ives and G. J. Janz, *Reference Electrodes*, Academic, New York, 1961.

I. M. Kolthoff and J. J. Lingane, *Polarography* Volumes 1 and 2, Wiley-Interscience, New York, 1952.

C. K. Mann and K. K. Barnes, *Electrochemical Reactions in Nonaqueous Systems*, Marcel Dekker, New York, 1970.

L. Meites, *Polarographic Techniques*, Wiley-Interscience, New York, 1965.

L. Meites and P. Zuman, *Handbook Series in Organic Electrochemistry*, Vols. I ff, CRC Press, 1978.

M. R. Rifi and F. H. Covitz, *Introduction to Organic Electrochemistry*, Marcel Dekker, New York, 1974.

S. D. Ross, M. Finkelstein, and E. F. Rudd, *Anodic Oxidation*, Academic, New York, 1975.

D. T. Sawer and J. L. Roberts, Jr., *Experimental Electrochemistry for Chemists*, Wiley, New York, 1974.

W. F. Smyth, *Polarography of Molecules of Biological Significance*, Academic, London, New York, 1979.

S. Swann and R. Alkire, *Bibliography of Electro-organic Syntheses* (1801–1975), Port City Press, Baltimore, Maryland, 1979.

A. P. Tomilov, S. G. Mairanoviski, M. Y. Fioshin, and V. A. Smirnov, *Electrochemistry of Organic Compounds*, Halsted, New York, 1972.

N. L. Weinberg, Ed., *Technique of Electroorganic Synthesis* Parts I and II (Techniques of Chemistry, Vol. 5), Wiley-Interscience, New York, 1975.

E. Yeager and A. J. Salkind, *Techniques of Electrochemistry*, Vol. 2, Wiley, New York, 1973.

P. Zuman, *Organic Polarographic Analysis*, Pergamon, New York, 1964.

REVIEWS AND OTHER SUGGESTED REFERENCES

M. M. Baizer, "Prospects for Further Industrial Applications of Organic Electrosynthesis," *J. Appl. Electrochem.*, **10**, 285 (1980).

O. R. Brown and J. A. Harrison, "Reaction of Cathodically Generated Radicals and Anions," *J. Electroanal. Chem.*, **21**, 387 (1969).

B. E. Conway, "Mechanistic Aspects of Anodic Oxidations," *Rev. Pure Appl. Chem.* (Australia), **18**, 105 (1968).

Z. Galus, H. Y. Lee, and R. N. Adams, "Triangular Wave Cyclic Voltammetry," *J. Electroanal. Chem.*, **5**, 17 (1963).

L. L. Miller, "Anodic Organic Chemistry" (about solvents), *Pure Appl. Chem.*, **51**, 2125 (1979).

F. D. Popp and H. P. Schultz, "Electrolytic Reduction of Org. Compounds," *Chem. Rev.*, **62**, 19 (1962).

A. P. Tomilov and M. Y. Fioshin, "Free Radical Reaction in the Electrolysis of Organic Compounds," *Russ. Chem. Rev.*, **32**, 30 (1963).

A. K. Vigh and B. E. Conway, "Electrode Kinetic Aspects of the Kolbe Reaction," *Chem. Rev.*, **67**, 623 (1967).

N. L. Weinberg and H. R. Weinberg, "Electrochemical Oxidations of Organic Compounds," *Chem. Rev.*, **68**, 449 (1968).

P. Zuman, "Preparative Methods in Elucidation of Organic Processes," *J. Polarogr. Soc.*, **13**, 53 (1967).

INDEX